Our Changing Menu

MR. PODELL,
THANK YOU FOR
YOUR INCREDIBLY
IMPORTANT SUPPORT!

OUR CHANGING MENU

Climate Change and the Foods We Love and Need

Mike P. Hoff
3-25-21

Michael P. Hoffmann, Carrie Koplinka-Loehr,
and Danielle L. Eiseman

COMSTOCK PUBLISHING ASSOCIATES
an imprint of
Cornell University Press

Ithaca and London

First published 2021 by Cornell University Press

Printed in the United States of America

ISBN 978-1-5017-5462-3 (paperback)
ISBN 978-1-5017-5464-7 (pdf)
ISBN 978-1-5017-5463-0 (epub)

Library of Congress Control Number: 2020950258

Illustrations by Lindsey Potoff

For support and love, Linda, Tara, Talya, and Matt
For inspiration, David Atkinson and John Oliver
—MPH

For Peg
—CKL

For my family,
but especially my uncle Marc for talking to me about rice
—DLE

CONTENTS

PREFACE

Why *Our Changing Menu*? Like everyone else we love and need food—that coffee at daybreak, the midday snack, or a scoop of ice cream on a hot summer eve. We are also deeply concerned about our changing climate. So we joined forces. Danielle, a lecturer at Cornell but once a chef, can spin a mouth-watering tale about the amazing world of cheeses faster than you can say "Parmigiano-Reggiano." Carrie, a writer, is passionate about sharing the stories that people in a changing world want to tell: a chef, a distiller, an olive scientist, an oysterwoman. Mike, who milked the family's cow twice a day in his youth, later traversed the vast farmlands of California and ultimately became a scientist. These and many more experiences gave him a deep respect for those who work the land and are more than ever challenged to supply the menu. The three of us seemed the right combination to tell this story.

You have probably heard plenty of worrisome stories about rising seas, intense storms, and fires but few about what climate change means to the many foods and beverages you love and need—the tea from India, olives from Spain, cocoa from West Africa, rice from California, or peaches from Georgia. Our goal in writing this book was to connect food with how the climate is changing it and how we can keep the menu stocked. Danielle, Mike, and a colleague published a survey in 2020 that suggests we are likely on the right track. Two-thirds of the respondents expressed a moderate or higher level of concern that climate change would affect their food choices. That survey used the broad term "food," whereas here we go deeper to look at specific foods we love and need.[1]

Writing *Our Changing Menu* was an exceptional opportunity to tell a science-based, factual story. Despite our personal biases we strove to use peer-reviewed literature, government reports, and other reliable sources to back up our story. We included over seven hundred citations. Not everyone will agree with every aspect of the story we tell, but be assured that the three of us and our technical reviewers worked hard to present the facts and include the best approaches to solving the enormous challenges facing our menu.

Telling this story was arduous yet an incredibly rewarding experience for us. Food is essential to life, the forms we enjoy are endless, it comes from every corner of the planet, and the path to our plate can be very complicated. We had to determine what to include and how to weave it together so that the consumer, restaurateur, policy maker, fisher, and retailer, to name a few, would gain from the story. Writing also created an opportunity to celebrate the richness, diversity, and flaws of our incredible—and changing—food system.

There are also no villains in this book. Many of us, including farmers, ranchers, employees of food businesses, governments, and nonprofits around the world, are striving to address climate change. We need to work together to address this grand challenge.

Once you have read *Our Changing Menu*, we hope you will be inspired to tell your story. Use your favorites—coffee, chocolate, potatoes, or your special comfort food—to convey the urgent need for action.

ACKNOWLEDGMENTS

We are deeply indebted to many people for the creation of this book, beginning with the experts who reviewed these pages. They not only saw when we strayed from fact but also noted when we might have strayed from reality. Given the diverse topics in the book, we engaged a wide range of reviewers, from climate scientists to a mixologist. Any errors that remain in the book are ours.

Our reviewers included Rima Al-Azar, Mike Baker, Chris Barrett, Brian Belcher, Gary Bergstrom, Brett Chedzoy, Michael Ciaramella, Jonathan Crane, Karl Czymmek, Art DeGaetano, Sanford Eigenbrode, Ben Faber, Dan Flynn, Brandon Fortenberry, Danny Fox, Tom Gallagher, Grant Gayman, Chris Gerling, Curt Gooch, Alexander Hristov, Jarra Jagne, Pat Johnson, Greg Jones, Kaylyn Kirkpatrick, Glen Koehler, Jordan LeBel, Johannes Lehmann, Karen Lewis, Bruce Linquist, Susan McCouch, Sara McDonald, Asha Miles, John Oliver, Greg Peck, Carlos Pérez, Carolyn Peterson, Per Pinstrup-Andersen, Paula Ribeiro Prist, Rick Rhodes, C. Alan Rotz, Earl Rutz, Ernie Shea, Adel Shirmohammadi, Tom Sleight, Timothy Spann, Margaret Smith, Mark Sorrells, Pat Sullivan, Norm Uphoff, Justine Vanden Heuvel, Paul Vossen, David Wolfe, Peter Wright, and Lew Ziska.

We add to this list David Hoffmann and Terry Kristensen, who provided "person on the street" reviews to help ensure our style and content were appropriate for the intended audience.

Many professionals who contribute to the world's food supply also agreed to be interviewed for this book. We thank Rima Al-Azar, Shannon Brock, Dan Flynn, Kyle Foley, Jesse Harriott, Jennifer Crist Kohn, Ken Martin, Taylan Morcol, John Oliver, Andres Padilla, Carlos Pérez, Pat O'Toole, Dave Rapaport, Allison Thomson, Tony Turkovich, Steve Train,

Krista Tripp, Tom Turini, and Colton Weinstein, all of whom had to come in from the fields or turn their attention from their work to share about their lives and perspectives.

Mike Koplinka-Loehr offered a listening ear and generous heart; Barbara Crooker allowed us to reprint a portion of *Ode to Olive Oil*; Kathleen Loehr lovingly shared her wisdom; the Lansing Writer's Group reviewed two years' worth of profiles; and Neel Inamdar described current challenges in US fisheries.

Cornell undergraduate students researched, fact checked, and wrote portions of some chapters. Maeve Anderson prepared drafts of the rice and chocolate topics, and Kalena Bonnier-Cirone drafted the vanilla and potatoes topics. Nancy Engel explored several topics early on.

We are grateful for the remarkable artwork of Lindsey Potoff (Cornell '22), who could take a boring bar graph and depict it visually for all to easily grasp. Her creativity was an enormous help in telling the changing menu story.

Kitty Liu, editorial director for Comstock Publishing at Cornell University Press, gave detailed editorial advice and responses to inquiries from the day we started discussing the book to the final days of going to press. Likewise, Alexis Siemon, acquisitions assistant and Mellon University Press Diversity Fellow at Cornell University Press, was exceptionally helpful.

Alison Fromme edited early drafts of several chapters. Kathleen Kearns, The Book Coach, read every word. She offered insights into the organization and structure of the story, checked for editorial compliance, and helped establish a consistent writing style. Her contributions were immensely helpful. She was an outstanding coach.

Cornell University's 2019 Podell Emeriti Award for Research and Scholarship and the Toward Sustainability Foundation provided critical financial support for the research, writing, and editing of *Our Changing Menu*.

Thank you.

Introduction

It's bad enough that climate change is melting glaciers and causing the seas to rise, but to many people the potential loss of coffee is downright scary. If not coffee, consider tea, spices, chocolate, seafood, rice, wheat, or whipped cream. The entire menu, including the before-dinner drink, salad, main course, and dessert, is all changing. *Our Changing Menu* is intended to provide a wake-up call by depicting the sweeping changes coming not only to the staples we depend on but also to many of our delectable favorites.

Despite the legitimate potential for doom and gloom when writing about food and climate change, this book is in part a celebration of the foods and beverages we enjoy: the aroma of coffee, the sting of a hot pepper, or the bitterness of a hoppy beer. It's also a refresher on the history of some delightful cuisines, where they come from, and their contributions to cultures and the world's economy. We don't intend this book to be comprehensive. We chose to focus on key examples of foods, beverages, and ingredients to which most everyone can relate.

The audience for this book is very broad—we all eat. Whether you're a connoisseur of fine wines, chef, baker, distiller, restaurateur, CEO of a food company, or someone who simply enjoys a good pizza or great drink, it's critical to know what is happening to our incredibly diverse and interwoven global food system. The emphasis is on the world's rich countries, but especially the United States—the "we" in *Our Changing Menu*—because cumulatively, we have contributed the most to climate change and are best positioned to do something about it.

This book delves into the backstories of the foods that appear at our local grocery or favorite restaurant. We include interviews with people who

When the climate dishes up change

Before Leña Brava opens its doors, burly, soft-spoken Andres Padilla has time for a conversation in the dining room. Within earshot, cooks chop vegetables and banter as they prep Baja-inspired cuisine. Padilla is creative culinary director for six restaurants in Chicago, each conceived and owned by Rick Bayless, the author of nine cookbooks and host of the public TV show *Mexico—One Plate at a Time*.

One of Padilla's jobs is to source achingly fresh ingredients so chefs can transform them into entrées that tug at your heart with whimsy and intrigue. He has a very local source: Leña Brava's rooftop and backyard gardens and greenhouses, which yield flowers, herbs, and hundreds of pounds of vegetables each year. He buys almost everything else he needs directly from farmers within 250 mi. (402 km) of the restaurant.

When Padilla first began sourcing, he could depend on ramps popping up nearby in late March, morels after that, and then peas. "We got accustomed to getting them at a certain time," he says. "Seasonality is a huge thing for us."

But lately, foods that appeared like clockwork years ago now arrive weeks ahead of schedule, may be available much longer, or disappear unexpectedly.

Mexican cuisine relies on tomatoes, but in 2018 Padilla asked himself, "Where did all the tomatoes go?" The chefs had to stop making their

work the land, fish the seas, and make our wine and beer. Millions of people around the world are on the front lines of climate change, and we introduce you to some of them.

The book is organized like a restaurant meal but opens with an overview of the world's food system, putting the menu in a global context. Just where does our food come from? We are accustomed to an extraordinary interconnected network that provides fresh produce year-round. Many people are surprised to learn that most grass-fed beef consumed in the United States comes from Australia, that a lot of tree nuts are from Vietnam, and that large quantities of fish caught by the US fishing fleet are processed in China, then shipped back to the United States. Food is also big business, providing 40% of global employment and 10% of consumer spending.

famous rooftop salsa when supplies failed and Padilla couldn't buy toma-toes elsewhere.

Similarly, he learned that warmer ocean temperatures are killing the kelp forests that sea urchins thrive on, threatening the supply of uni, the roe-producing gonads of sea urchins. "I've had to say, 'Sorry, chefs, no uni today,'" recounts Padilla. "I would try to order it, and the divers weren't having any luck."

Padilla wants to buy from sustainable fisheries but verifying a long chain of sources and distributors can require research for which he doesn't have time. He consistently opts for fish raised in farms he has visited because they are dependable and he knows exactly from where the fish are coming.

Two other items on the menu are rare mescals and tequila. Although both are distilled from the crushed hearts of agave plants, tequila is made from 100% blue agave, which can be cultivated. On the other hand, each unique mezcal is made from wild agave plants, some very rare and re-gion specific. "One kind, *tobolá*, grows up high between crevices of rocks, and people have to climb mountains to harvest it," says Padilla. The plants are sensitive to human activity and changes in the climate. "If we continue to harvest at the rate we're harvesting without replanting to the wild, some types of agave may go extinct."

With a global view of the food system as a foundation, we shift to the broad challenge posed by a rapidly changing climate. Why is it changing? Where do greenhouse gases come from? What does the future hold? We live in a very thin layer on the surface of the planet, much like a peel on an apple, and it is warming because of the greenhouse gases we're pumping into it. And as we continue to warm this thin layer, glaciers melt, seas rise, and storms intensify; the climate changes. Be prepared, as this topic can be depressing and overwhelming, but it's an essential starting point.

All of these changes affect the plants we depend on for life. Plants need air, water, the right temperature, soil, and sunlight. The air now has more of the greenhouse gas carbon dioxide, which means that most plants will grow faster and bigger, but any benefit will be offset by stress from increas-ing heat and drought. More carbon dioxide also means less nutritious crops in the future and tougher pests to control. Precipitation for our crops is becoming less dependable in many regions. Rising nighttime temperatures are reducing yields of some crops. At the same time, some of the shifts may mean new crops can be grown in more northerly regions, benefiting their economies.

Strikingly, it's not just our favorite foods that are changing but also perfumes, flavors, pet foods, pharmaceuticals, cosmetics, medicinal herbs, clothing, and soaps. All these items, from the flowers used in perfume to the cotton in our clothing, depend on plants.

With this as background, we look closely at a meal, starting with alcoholic beverages. These beverages have been around for thousands of years, are important to many cultures, and are major contributors to today's economy. Beer, wine, and spirits depend on grains, grapes, other fruits, water, herbs, and spices (gins tap about 150 botanicals). That's all changing. Warmer winters in Belgium are threatening the production of some Lambic beers. Droughts are affecting not only the availability of hops but also the quality of water used in brewing beers. High temperatures alter the aromatic compounds and sugar levels in wine grapes and also increase the "angel's share" in bourbon production—the amount lost to diffusion through the oak barrels. Despite these challenges, brewers, vintners, distillers, and those who supply the essential ingredients are working hard to adapt, reduce their impacts on the climate, and keep these delightful beverages flowing.

The delicious and diverse salad is next on the menu. Increasing temperatures and extreme weather are affecting the production of salad ingredients, including greens, fruits, herbs, spices, and avocados. The United States is the largest importer of avocados worldwide, consuming about 4,000 tons (3,600 metric tons) of them on Super Bowl Sunday, mostly as guacamole. Olives are under stress in the Mediterranean because of higher temperatures and increasing droughts. Growers are adopting new ways to reduce stress on orchards, and some production is already shifting to regions where climate change is less challenging.

Following salads? The main course: pork, lamb, fowl, fish, and beef, which we eat a lot of in the United States. Meat has been part of the human diet for over two million years and is an essential component of diets for millions of people in developing countries. But the higher temperatures and droughts are affecting the animals we eat. At the same time, greenhouse gas emissions from the animal sector are significant and are starting to be addressed through improving animals' diets and grazing systems. This book encourages people in the United States and other rich countries to shift to a more plant-based diet and to view meat, and in particular beef, as a delicacy rather than a staple.

Chicken, another main course option, is the most widely consumed meat in the world. Along with chicken, the United States produced over one hundred billion eggs in 2017. Chicken and egg production in the United States is somewhat protected from climate change, since most fowl are enclosed in houses, but intensifying storms and heat waves have caused losses globally.

Fish and other aquatic foods can round out the main course choices. Seafood is big business, with the US fish industry harvesting 5 million tons (4.5 million metric tons) per year. The United States imports an additional 3 million tons (2.7 million metric tons). The most common seafood eaten in the United States is shrimp, usually imported from Thailand or China.

Because of climate change, the vast oceans are warming and becoming more acidic, with profound implications to this critically important source of human food. In some regions, phytoplankton—small floating plants that form the basis of the food chain—are declining, and oysters and clams, which depend on shell formation, are threatened by the increasingly acidic conditions.

What is needed to keep the menu supplied with this incredibly important food source? Given the scale of the challenge it is imperative to reduce greenhouse gas emissions worldwide, manage wild fisheries more wisely, and expand sustainable aquaculture.

Starches, grains, and other side dishes accompany the main course. Many of the grains are ubiquitous, such as rice, a staple for 3.5 billion people worldwide. Wheat follows closely behind rice in global importance, and in the United States, the potato is the leading vegetable crop. All are under increasing risk due to climate change, but in response, scientists are developing more climate-resilient varieties.

We ultimately arrive at dessert and coffee. Many flavors and key dessert ingredients, such as nutmeg, milk, maple syrup, vanilla, coconut, and sugar, are changing. Overheated cows give less milk; intensifying storms in Madagascar destroy vanilla farms; and a lot of coconut is grown near rising seas. Our beloved coffee is also changing as temperatures warm and rain falls at new times of the year. Here again, researchers are developing

Keeping current: Our Changing Menu website

We have created an online presence, Our Changing Menu (http://our changingmenu.com/), to complement this book. Because the foods we love and need are changing so rapidly, we believe it's important to keep information current. This website includes a searchable database of foods, beverages, and their ingredients so that consumers, chefs, retailers, food processors, and others in the food business can learn what is happening to their essential or favorite ingredients from around the world. Each item includes the nature of the impact, such as drought, floods, or high temperatures, its severity—currently and what is projected for the future—and where the impact is occurring (e.g., vanilla in Madagascar, spices in India, wine in California). Images, graphics, and sources for the science-based information are included. An interactive component allows users to share their experiences of changes in foods due to climate change.

new and hardier varieties and farmers are growing crops under shade trees to reduce heat and water stress.

After illustrating how the menu is changing, we switch to what farmers, businesses, and scientists are doing to save it. Farmers, the stewards of the land, are using water efficiently, keeping the soil healthy, adopting tougher crop varieties, and producing renewable energy with solar and wind. Many food businesses are also reducing their risks from climate change by assessing threats along their supply chains and supporting growers around the world to build resiliency.

The scientific community is also responding by developing climate-resilient farming tactics, crops more tolerant to heat and drought, and improved predictions of severe weather, to name a few. Unfortunately, federal support for research and development in the United States has declined. The government spends about $4 billion annually on agriculture and food research. To give this some perspective, people in the United States spend over $43 billion per year on video games and accessories. With the multitude of new challenges facing the food system, science—the pursuit of knowledge—needs to be valued and supported like never before.

We wrap up by describing what we can do. By reading this book you will be better informed about the causes of climate change, how each of us affects the climate, and how it affects us. You'll have a good grasp of the scale of food waste and what to do about it, why a move to a plant-based diet is very helpful, and the need to appreciate and support those who supply our menu. This book will prepare you for action.

Having read *Our Changing Menu*, you will be able to share the climate change and food story with friends, relatives, neighbors, and those in positions of influence—business leaders, elected officials, and policy makers. Raise this highly relevant issue to a new level—the effects on the economy, jobs, and global social unrest, let alone the foods we love and need. If we raise our voices, we may be able to start a social movement around food and climate change and help bring about the sweeping changes needed. We all eat, and it's all changing.

We close with a reminder that none of us are alone. Thousands of people are fighting this fight. Everyone needs to get involved and consider this book a gift, a way forward. Now be courageous and lead.

BACKGROUND

SETTING THE TABLE

To set the stage for this book, we start with a description of where the food we love and need comes from and how it gets to our table. Not enough of us know. The fresh vegetables we enjoy may come from a local farmers market, the grapes from California, tree nuts from Vietnam, coffee from Brazil, spices from India, and fish from the Bering Sea, to name a few. This global interconnected and interdependent food system that feeds us also provides 40% of global employment and accounts for 10% of consumer spending—a $5 trillion business. But it faces increasing risks from a changing climate.

Next we describe why and how the climate is changing and how this is affecting the world around us with record-breaking heat, more intense storms, melting ice, warming oceans, and rising seas. We then provide an overview of how these many changes are affecting plants, the basis of life. Plants—from the giant redwoods to wheat—require the right temperatures, water, soil, air, and sunlight. All of these requirements except sunlight are changing, with subtle to profound implications. In these beginning chapters we discuss the impacts on plants on land and in the oceans and wrap up with an overview of how pests, pollinators, and the food supply chain are being affected by a changing climate.

Our Food Supply

From Land and Sea to the Menu

In today's globally interconnected food system, many people are privileged enough to dine on food from hundreds of locales around the world, whether it's fresh kiwis from Peru, mangoes from Mexico, saffron from Iran, or wine from Italy. An extraordinary system—farms, facilities for transport, storage, and processing, and wholesale and retail suppliers— helps ensure that many who live in the United States or other rich countries have a full menu of foods and drinks from which to choose.

The food you eat has often traveled great distances and gone through a number of processes before it reaches your table. As you sit down to quiet your growling stomach with a pasta dish, pizza, or a chicken casserole, you likely will not know from where the many ingredients came. The farmer in Kansas who harvests the wheat for your sandwich bread, the families on small farms in Costa Rica who pick the beans for the coffee you savor, and the Tanzanian farmer growing sesame for the oil in your salad dressing are all part of this amazing global food system.

Not all food is from distant places. In 2012, about 8% of US farms were marketing their products locally.[1] However, for many who rely on the grocery store to fill the refrigerator, much of their food must travel great distances before it reaches the store's aisles. Although the United States produces most of its own food and exports a lot, increasing economic growth has led to greater imports of fruits, vegetables, and tropical products.[2] Some examples include vegetables from Mexico and Canada, fresh cheese from Italy and France, and tree nuts from Vietnam.[3]

Our food starts on a farm, on a ranch, or in the sea. After harvest, it might travel to a local market or a processing facility where it can be transformed into a food product and packaged. The processed food then travels to a wholesaler and next to a grocery store or retailer, where the consumer can purchase it. Some foods take a circuitous route. For example, large quantities of fish caught by US fishing fleets go to China for processing and then are shipped back to the United States.[4]

The United States has imported and exported food since colonial days, but over time a small, localized, and fragmented system has become a global, more efficient one. In the past few decades, companies began importing foods that cost much less to bring in than to grow domestically while exporting other products to increase revenue.[5] This shift increased the distance food traveled by 20%.[6]

The food system has also consolidated in recent decades. Four agribusinesses currently control 90% of the global grain trade.[7] Nestlé, which started out making condensed milk in Switzerland in 1866, is now one of the world's largest food and beverage manufacturers. The company has 413 factories in 85 countries and sells products in 190 countries.[8] Consolidation allows for more efficient production and cheaper costs, but it also means a few large global companies with extensive reach and thousands of employees—such as Cargill, Walmart, ConAgra, Archer Daniels Midland (ADM), and Unilever—dominate the food system.

Though it is difficult to determine the economic value of the complex global food and agribusiness industry, McKinsey & Company estimates it is worth $5 trillion. Overall, the industry provides 40% of global employment and accounts for 10% of consumer spending.[9] These numbers vary within countries and specific parts of the supply chain. In the United States, agriculture and food-related industries provide about 22 million full- and part-time jobs, 11% of total employment. They generate over $1 trillion, about 5% of the GDP. At 13%, food is the third-largest US household expense, behind housing and transportation.[10]

The food supply chain encompasses four distinct activities: production, processing and packaging, distribution and retailing, and consumption. It can be broken down further into activities and inputs, such as land, labor, crop maintenance, storage, transportation, regulations, subsidies, and marketing. Historically, farming was the dominant economic activity in the food production chain but it has now been replaced by value-added activities such as processing and packaging.[11]

As the global population has rapidly grown, so, too, has our global food production, which tripled between 1961 and 2011. Food production practices have become more efficient, and farmers can generally produce more on the same or, in some cases, less land. On the same-sized plot of land used in 1961, three times as much cereal grain can be produced today.[12] Higher yields of commodities per unit of land have brought the cost of many desirable foods down, benefiting consumers. Globalization and trade agreements have also generally helped to provide a greater diversity of foods to consumers across the globe.

Looking to the future, the world faces three grand challenges in maintaining its food supply. The first is a population increasing in size and purchasing power and becoming more urbanized, requiring more transport, storage, and preservation of food. Next is preserving agricultural lands to provide enough food while ensuring the livelihoods of food producers. And last, the world's poorest people must be nourished.[13] We need to achieve all this even as unprecedented climate changes make it more difficult to maintain or increase food production. These challenges are daunting, but researchers, governments, private industries, nonprofit organizations, and many others are exploring ways to meet them.

Summing It Up

Most people in developed countries can access foods through an incredible global system from nearly every corner of the earth. This system has thousands of players, and in the United States alone provides twenty-two million full- and part-time jobs. Production efficiency has increased markedly over the past few decades, but grand challenges remain—an increasing global population, the need to preserve land for agriculture, and feeding the world's poor.

This global system is under intensifying stress, and the next chapter provides a primer on a major source of that stress—the changing climate.

Our Changing Climate

We live in a small space. Though the earth can seem vast, as can the endless skies above, in fact we exist in a wafer-thin layer of the atmosphere averaging about 7 mi. (11 km) above sea level.[1] That's it. A very thin layer—think apple peel—on a planet that has been not too hot, not too cold, but just right for the past ten thousand years.[2] With these relatively stable conditions, we have built great cities, created extraordinary communication and transportation systems, and populated almost every corner of the earth, about eight billion tenants and growing. And now, we are jeopardizing that thin layer—our home, the birthplace of every human being who has ever lived.

In that thin layer of atmosphere, weather occurs: the snow falls, fogs swirl, clouds form, and the wind blows. That dynamic layer, which is called the troposphere, consists mostly of nitrogen (78%) and oxygen (21%), along with greenhouse gases that human activity is increasing, such as methane (CH_4), nitrous oxide (NO_2), and, most important and most persistent, carbon dioxide (CO_2).

If these critical greenhouse gases were absent, the earth's average annual temperature would be well below freezing instead of relatively warm (59°F, 15°C), and human life would not exist. However, we have been pumping enormous quantities of these gases into that thin layer, especially with the start of the industrial revolution in the late 1800s. By burning more coal, oil, and natural gas and generating more carbon dioxide, we are making that thin blanket denser. This is somewhat like adding a heavier blanket on our bed to hold the warmth radiating from our body. We are warming the atmosphere, a process referred to as global warming, which is causing the climate to change. And the changes include increases

We live in a very thin layer on the surface of the planet—much like the peel of an apple.

in extreme weather events, melting of glaciers and ice sheets, and sea level rise. Humans are causing the climate to change—it's anthropogenic.

Warming is relatively simple. When direct sunlight (solar energy) hits a snow-covered field or ice-covered continent, most reflects back into space. However, when sunlight hits other surfaces—a highway, a dark roof, or a vast desert—the surfaces absorb the energy and radiate it back into the atmosphere as heat. Greenhouse gases absorb much of that heat, trapping it in the troposphere. Even though these gases occur in exceedingly small amounts, they have a huge warming effect. Since the beginning of the twentieth century, atmospheric carbon dioxide has increased to over 400 parts per million (ppm), over 40% higher than it should be. The result, an atmosphere that has warmed 1.8°F (1°C). And since some of the carbon dioxide remains in the atmosphere for hundreds to thousands of years, we face a serious long-term challenge. If we continue on our current trajectory, carbon dioxide could exceed 900 ppm by the end of the century. Add in the other greenhouse gases and the result is a potential 9°F (5°C) of warming.[3] That is dangerously hot.

Many necessities and conveniences in our lives, such as lighting, heating and cooling our homes, travel for fun or work, and what we eat, generate greenhouse gases. There were over one billion cars on the planet in 2018, and the number is expected to double by 2040.[4] We also love to fly,

Climate change not slowed by COVID-19

During the early months of the COVID-19 pandemic in 2020, global carbon dioxide emissions dropped 17% compared to the previous year. The decrease was dramatic and the result of a shutdown of large components of the global economy. Unfortunately, it was only temporary, and just a few months later, global carbon dioxide emissions were only about 5% below those of the previous year.[a] With 95% of the emissions still occurring, this drop in emissions was not nearly enough to change the predictions that the year 2020 would be one of the hottest on record or halt the march of climate change.[b] Some people considered COVID-19 to be a test of international cooperation that would help the world prepare for the much larger crisis of climate change on the horizon.[c] Unfortunately, all signs indicate that the world did not pass the test.

---

[a] Corinne Le Quéré et al., "Temporary Reduction in Daily Global CO_2 Emissions during the COVID-19 Forced Confinement," *Nature Climate Change* (May 19, 2020): https://doi .org/10.1038/s41558-020-0797-x; Corinne Le Quéré et al., "Supplementary Data to: Le Quéré et al. (2020), Temporary Reduction in Daily Global CO_2 Emissions during the COVID-19 Forced Confinement," Global Carbon Project (2020): https://doi.org/10.18160 /RQDW-BTJU.

[b] Maanvi Singh, "A Summer Unlike Any Other: Heatwaves and Covid-19 Are a Deadly Combination," *Guardian*, May 30, 2020, http://www.theguardian.com/us-news/2020 /may/30/coronavirus-heatwaves-health-summer-us-cities.

[c] Oscar Serpell, "Climate Change Take-Aways from the Global COVID Stress-Test," April 29, 2020, https://kleinmanenergy.upenn.edu/blog/2020/04/29/climate-change-take -aways-global-covid-stress-test.

with thirty-nine million flights globally in 2019.[5] In the United States, the primary sources of greenhouse gases are transportation, the production of electricity, industry, commercial and residential heating, and agriculture.[6]

The Evidence Is All Around Us

Physical evidence going back over thousands of years shows an obvious relationship between changes in carbon dioxide levels and the temperature of the atmosphere.[7] When carbon dioxide levels in the atmosphere

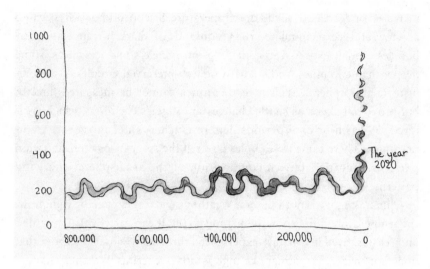

Atmospheric carbon dioxide levels (ppm) over the past 800,000 years and projection to 2100 (dashed). (Recorded data from "Graphic: The Relentless Rise of Carbon Dioxide," National Aeronautics and Space Administration, https://climate.nasa.gov/climate_resources/24/graphic-the-relentless-rise-of-carbon-dioxide. Projection from K. J. Hayhoe et al., "Climate Models, Scenarios, and Projections," 138.)

Sources of greenhouse gases in the United States by economic sector. (Data from "Sources of Greenhouse Gas Emissions," US Environmental Protection Agency, https://www.epa.gov/ghgemissions/sources-greenhouse-gas-emissions.)

increased or decreased, so did the temperature. Scientists know this because as snow and ice accumulated over thousands of years, it trapped tiny air bubbles. Today, experts can drill deep into glaciers and ice sheets, some deeper than two miles, and study those well-preserved samples of the pre-historic atmosphere. Ancient temperatures cannot be measured directly, but proxy data such as subtle changes in oxygen's chemistry, which is altered by warm or cool periods, lets researchers infer those early temperatures. These same ice samples also tell the story of past precipitation patterns, volcanic activity, composition of the atmosphere, and wind patterns.

Climate change is not weather. Weather is what is happening right now: the temperature, rain, snow, wind, and so on. It can vary by the hour, day, and season. Even if you have experienced three cold winters in a row, that does not mean that climate change is not occurring. Weather changes quickly, but climate change is long-term—it happens over decades.

It's getting warmer. Globally, 2015 to 2019 were the hottest years since observations started in the nineteenth century, and records continue to be broken.[8] Consequently, the world's ice is declining quickly.

Glaciers are rapidly melting, worldwide. For example, the glaciers in Glacier National Park in Montana are likely to be gone by 2030.[9] Not only are glaciers phenomenal to see, they also are incredibly important for

Ice melts when it warms.

agriculture, drinking water, and hydroelectric power. In the Himalayas, some estimates indicate that much of the glacial ice will be gone by the end of the century, affecting tens of millions of people.[10]

Glaciers are disappearing and so is the sea ice covering the Arctic Ocean. It's retreating because over the past several decades the region has warmed twice as quickly as the rest of the planet. In September 2019, summer sea ice covered about 811,000 mi.^{2} (2,100,000 km^2), less than the 1981 to 2010 average.[11] That additional open water is almost the size of Alaska and California combined. With such a large swath of ocean absorbing 90% of the sun's warmth, more floating ice melts, which opens up more ocean—you get the picture. Although the sea ice is melting, it does not contribute to sea-level rise, since it's already floating in the sea.

Ice sheets are a different story. They are large masses of glacial ice that cover thousands of square miles of land, and are up to 3 mi. (4.8 km) thick. Together, the Antarctic and Greenland ice sheets contain over 99% of the world's freshwater, and both are melting. The Antarctic ice sheet is about the size of the United States and Mexico combined, and the Greenland ice sheet is about three times the size of Texas.[12] Since 2002, the Greenland ice sheet has been losing 280 billion tons (254 billion metric tons) of ice per year.[13]

As glaciers and ice sheets melt, trickles of water become rivers that eventually pour into the oceans. Warm water takes up more space than cooler water, and these meltwaters and warming oceans contribute in about equal parts to sea-level rise. Since 1900, the global average sea level has risen about 7 to 8 in. (18 to 20 cm), nearly half of that since 1993. Sea levels will likely rise 1 to 4 ft. (0.3 to 1.2 m) by century's end, possibly 8 ft. (2.4 m).[14]

In the United States, people living near the coast will be at increasing risk.[15] Rising seas will inundate some island nations in the Pacific and hit countries such as Bangladesh, with its millions of coastal residents, especially hard. Saltwater intrusion, partly the result of rising seas, is already threatening freshwater supplies in Florida, as well as food production in the Mekong Delta, Vietnam's rice bowl.[16]

The oceans absorb about one-third of the carbon dioxide currently emitted into the atmosphere by human activity, and the carbonic acid that results has increased ocean acidity by about 30% over the past two hundred years.[17] The effect is not uniform, as cold water absorbs more carbon

A billion tons (one gigaton) of ice compared with the Empire State Building.

dioxide than warm water, but the overall acidification rate is unprecedented; nothing like it has occurred in the past sixty-six million years. This increased acidity interferes with shell formation in lobsters, clams, oysters, corals, and plankton, the foundation of the ocean's food web.

Meanwhile, torrential downpours have increased in intensity and frequency; heat waves are becoming more common and intense, and forest fires burn longer and with greater ferocity, especially in the western United States and Alaska. Earlier spring snowmelt and less snowpack have already threatened water supplies in the western United States and elsewhere.[18] In a nutshell, climate change is causing more extreme events today, and these will get worse in coming decades.

Because of temperature, precipitation, and other changes in the physical world, the biological world is changing too. One of the best examples is the explosion of bark beetle populations in the western United States and Canada. Warmer summers increase their growth rate, and warmer winters permit greater survival. Drier conditions have also stressed trees, allowing beetles to attack them more successfully. The result is millions of acres of dead and damaged trees across vast regions. The damage by bark beetles is expected to grow in coming years.[19] Unfortunately, even California's nearly indestructible giant sequoias are succumbing to beetle attacks.[20]

And it's not just bark beetles. Certain ticks, mosquitoes, and plants, including some weed species, are now surviving in more northerly ranges. The evidence of climate change is all around us.

Birds are in trouble too. Of 604 North American species of birds surveyed, almost two-thirds are considered moderately to highly vulnerable to climate change. The degree of vulnerability depends on habitat and breeding season, but overall, the rapid pace and scale of climate change makes it an existential threat to birds.[21] They are the "canaries in the coal mine."

Millions of species around the world—animals, plants, and microbes—provide us with food, shelter, medicines, and much more. They also help form soils essential for agriculture and regulate climate at the local and

The much-loved puffin is in serious decline across the eastern Bering Sea and elsewhere because warming oceans are affecting its food sources. (Timothy Jones et al., "Unusual Mortality of Tufted Puffins [*Fratercula cirrhata*] in the Eastern Bering Sea," *PLOS ONE* 14, no. 5 [May 29, 2019]:1–23, https://doi .org/10.1371/journal.pone.0216532.)

global scale. However, one million species may already face extinction for a variety of reasons, including climate change.[22]

Changes to the Earth's Climate—Are They Natural?

The earth's climate has changed many times owing to natural causes and continues to change in part because of these same causes:

- The earth's orbit around the sun is elliptical, not a perfect circle, so at times it is closer to the sun, which has an increased warming effect, but the earth is not getting closer at the present time.
- The earth also tilts and wobbles periodically about every 41,000 and 23,000 years, respectively, causing changes in the climate. This too is not occurring now.
- Volcanic activity can cool the climate by expelling reflective material into the atmosphere, a one- to two-year effect that has little impact on the long-term warming trend.
- Volcanoes can also add carbon dioxide to the atmosphere, but this amount is very small compared with human contributions, so it does not have enough of an effect on the climate to explain what is now being observed.
- The amount of solar energy the sun emits varies slightly and can influence the climate, but an increase of 0.1%, which is expected every eleven years, does not explain the level of warming that has occurred on the planet.
- Every three to ten years, natural shifts in trade winds change the flow of warm waters in the tropical Pacific Ocean, affecting global weather patterns. This warm-water phase is referred to as El Niño and the cool phase as La Niña. These cycles are relatively well understood but don't last long enough to explain the long-term warming we are experiencing.

In sum, natural influence is too small and too slow to explain the rapid changes under way today.

The Future—What Does It Hold?

Climatologists use models to help predict future conditions, just like a Global Positioning System (GPS) uses data plus a simple formula to estimate drive time to a destination (distance/speed = time). Climate models that project conditions decades from now are far more complicated and may include data related to incoming solar energy, outgoing energy (heat), greenhouse gas concentrations, aerosols in the atmosphere, land cover, humidity, and ocean temperatures. Scientists have confidence in the models because they are based on well-established laws of physics and numerous observations. Models are also often tested in a process called hindcasting. If a model comes up with numbers that accurately match recorded past conditions, such as temperature, then scientists can have some confidence in its forecasts of the future. Many different models exist, and often analysts bring various simulations together to show the range of predictions, somewhat like the way meteorologists predict multiple possible paths for a hurricane. Climate models are generally quite accurate at predicting a warming climate due to increases in greenhouse gases.[23]

The Scary Part

The historic 2015 Paris Agreement called for preventing the global average temperature to exceed 3.6°F (2°C). A subsequent 2018 report focused on strengthening efforts to keep it below 2.7°F (1.5°C).[24] These goals were set because climate change poses a potentially irreversible threat to natural systems and human societies. Analysts also predict scenarios in which impacts could pile on top of each other—one impact joining forces with another.

One worrisome scenario is several climate events happening at the same time or in rapid sequence, creating an additive effect. Examples include the combination of high temperatures, drought, and wildfires followed by massive mudslides on the denuded lands, something we have already seen in the western United States. Or a drought could occur simultaneously in major agricultural areas around the globe, putting at risk the food supply for billions of people. On a regional scale, researchers predict a 99% chance of a megadrought (over thirty years long) hitting the western and southwestern United States before the end of the century.[25] Scientists have more recently reported that the years 2000 to 2018 were the second driest for the region since the year 800 CE. Global warming has pushed a moderate drought into a megadrought.[26]

When you chop down a tree, a point comes at which one last swing of the ax will make the tree fall, irretrievably. That's a tipping point, the moment we exceed a threshold and enter an irreversible state. With declining ice sheets, more of the earth's surface is exposed and absorbs more solar energy. This adds to the warming of the atmosphere and oceans, which leads to more melting of ice sheets—an irreversible process once the tipping point is passed. Another example would be the thawing of the permafrost in the Northern Hemisphere. As it thaws, both methane and carbon dioxide will be released, increasing the greenhouse effect, further warming the atmosphere, and melting more permafrost. These and several other tipping points are a growing and serious concern, especially if one nudges another in a domino effect.[27]

Current climate models are good at predicting what lies ahead, and what they tell us is scary enough, but they can't predict every potential catastrophic and irreversible twist and turn. Scientists and policy makers agree that warming beyond 2.7°F (1.5°C) is very dangerous, and we will likely reach that threshold between 2030 and 2052.[28] Many experts agree that unabated climate change is an existential threat to civilization as we know it today.[29]

Summing It Up

The science is solid. Our climate is changing rapidly, and humans are the cause. And the evidence is all around us: increasing greenhouse gases are warming the atmosphere, ice is melting worldwide, storms are intensifying, seas are rising, and many species are at risk from the unprecedented changes. We simply have no choice but to meet this grand challenge head-on. Humans are inspiringly innovative and courageous when our backs are against the wall. To save our favorite foods, our favorite beverages, we all need to act. The words Abraham Lincoln spoke in 1862 in a very different time and context apply to the climate change challenge we face today: "The dogmas of the quiet past are inadequate to the stormy present. The occasion is piled high with difficulty, and we must rise with the occasion. As our case is new, so we must think anew and act anew."[30]

With an appreciation for the cause of climate change and how it is impacting the world around us, the focus now shifts to how the menu is changing.

Climate Change

How It Is Fundamentally Altering the Menu

A plant, whether wheat or a giant redwood, needs water, air, sunlight, the right temperature, and soil. Despite their many variations, the plants we depend on for food have these basic needs, and the animals we elect to eat have similar ones.

Plants may consist of up to 90% water, which has several roles, including moving nutrients throughout the plant and providing the turgor, or water pressure, that keeps stems and leaves upright. Water also keeps the plant cool by evaporating through small pores in the leaves. During photosynthesis, the water molecule (H_2O) splits, providing oxygen to the plant

What a plant needs for life—water, air, soil, the right temperature, and sunlight.

as well as to humans and animals. Hydrogen is also released. Plants use a lot of water, sometimes equal to their own weight per day, and how much they need varies with the wind, temperature, and humidity.

Air is the source of carbon dioxide, which provides the carbon that combines with the hydrogen from water to make sugars, the plant's energy currency and ultimately all the earth's food. Sunlight, or solar radiation, provides the energy that drives these chemical processes. Fortunately, solar radiation is one thing that is not changing with a changing climate. The temperature needs to be just right, about 77°F (25°C), for optimal photosynthesis to occur. If the air is on the warm side, everything speeds up; if cool, it slows down. If too hot, the plant may lose water faster than it can take it in; turgor is lost, and the plant wilts. If too cold, the water within the plant freezes, destroying cells. Soil is the source of both water and the essential nutrients critical for healthy plant growth. Most plants also depend upon the soil as an anchor for their roots.

Our description simplifies a complex system and exceptions exist, but these essentials allow a plant to grow, develop, and reproduce. These necessities—water, air, temperature, and soil—are changing as the climate changes, directly or indirectly affecting the food system on which we depend. Some of the changes are beneficial, but most are detrimental. The result? Our menu is fundamentally changing, with profound implications across the globe.

The Water

Let's start by putting water in perspective. Ninety-seven percent of water occurs in the oceans as saltwater; only 3% is fresh, and of that, 99.7% occurs as icecaps, glaciers, or groundwater. The remaining 0.3% is surface water—lakes, wetlands, and rivers. Most of our crop plants depend on this tiny fraction, this surface water, but climate change is bringing about fundamental changes in the entire global water cycle. A warmer climate causes more evaporation from land and bodies of water, and a warmer atmosphere holds more water—about 7% more water for every 1.8°F (1°C) rise in temperature.[1] And a warmer climate will affect the amount, timing, distribution, form, and quality of water to grow, process, and transport our food.

How much precipitation (rain, snow, sleet, hail) falls is changing around the globe. Europe and North America are experiencing increases, while

The global water cycle is changing, from the amount of evaporation from plants and surface waters to the availability of groundwater.

in the Mediterranean, most of Africa, and southern Asia, annual precipitation is decreasing. Experts project that globally, dry areas will get drier and wet areas wetter.[2] In the United States, precipitation has increased in the fall but not during winter months. Parts of the Southwest and Southeast have gotten drier, while the Northeast, Great Plains, and Midwest have gotten wetter.[3] Shifts in precipitation patterns will result in some regions, such as the inland northwestern United States, having more frequent water shortages and others, such as the northeastern United States, having adequate water for agriculture if managed appropriately.[4] Regional droughts are likely to increase in frequency and duration in coming decades and reduce crop production.[5]

Agriculture in regions that receive less precipitation in coming years will, in many cases, rely more on groundwater for irrigation. However, the withdrawal rates of many of the world's major aquifers are exceeding the rate of recharge, and some will last only a few more decades.[6] The Ogallala Aquifer is the single greatest source of groundwater in North America, providing irrigation water to the Great Plains states and helping to produce one-fifth of US corn, wheat, and beef cattle. It is already greatly overdrawn and not being adequately replenished.[7]

Those precious almonds

Adding toasted almond slivers to your salad? They're packed with protein, fiber, "good" fat, essential minerals, and vitamins.[a] Healthy and versatile, almonds can be eaten raw or cooked and are often ground to make flour, milk, oil, butter, and paste (marzipan). No wonder more than 2 lb. (0.9 kg) of almonds were eaten per person in the United States in 2017.[b]

California grows about 80% of the world's almonds, contributing a whopping 1.1 million tons (0.99 million metric tons).[c] Almond trees stretch for miles in the San Joaquin Valley and normally bloom from February to March. Because the winter chilling (dormant) period has begun to shorten as the climate warms, some experts predict that by 2030 almond yields will decrease by 10%. One solution is to switch to varieties that tolerate warmer winters.[d]

Almonds also need a lot of water, more than walnuts or pistachios. With the doubling of grove acreage in the past two decades, almonds are quickly becoming California's most extensive irrigated crop.[e] Fortunately, according to the CEO of the Almond Board of California, almost all the state's almond farmers have adopted precision irrigation technology, watering roots instead of entire orchards.[f] To help recharge groundwater, some growers flood orchards during winter rainstorms. Nonetheless, if multiyear droughts recur and the snowpack in the Sierra Nevadas continues to dwindle in the coming decades, conserving water will be even more critical.

To complicate matters, it is not clear how shifts in precipitation patterns caused by climate change will alter the recharge of aquifers.[8] Changes in precipitation patterns also affect water levels in lakes, reservoirs, and rivers, all of which can provide irrigation for our crops. Finally, hotter and drier conditions around the world have increased the incidence and severity of fires across large swaths of land, taking a major human toll and causing enormous economic losses, including those to agriculture.[9]

The United States is experiencing more downpours, and that trend is expected to increase in the coming decades, especially in the Northeast.

In addition to conserving water, the California almond industry is utilizing almond by-products and reducing wastes. Almond shells and hulls can become livestock feed and bedding, fuel for creating electricity, and mulch that boosts nutrients, yields, and soil carbon.[9] The US Department of Agriculture and several leading companies are developing hull-based bioplastics, biofuels, and even a sugar substitute for making beer and hard cider.[h]

---

[a] "Nuts, Almonds," Food Data Central, National Nutrient Database, US Department of Agriculture (USDA) Agricultural Research Service, April 1, 2019, https://ndb.nal.usda.gov/ndb/foods/show/3635?format=Full.

[b] "Almonds," Agricultural Marketing Resource Center, revised October 2018, https://www.agmrc.org/commodities-products/nuts/almonds.

[c] "Tree Nuts: World Markets and Trade," USDA Foreign Agricultural Service, October 2018, https://downloads.usda.library.cornell.edu/usda-esmis/files/tm70mv16z/1g05fg07r/r207ts47n/TreeNuts.pdf.

[d] David B. Lobell and Christopher B. Field, "California Perennial Crops in a Changing Climate," Climatic Change 109 (December 2011):330, https://link.springer.com/article/10.1007/s10584-011-0303-6.

[e] Julian Fulton, Michael Norton, and Fraser Shilling, "Water-Indexed Benefits and Impacts of California Almonds," Ecological Indicators 96 (January 1, 2019):715, https://doi.org/10.1016/j.ecolind.2017.12.063.

[f] Daniel Beaulieu, "Growing Almonds with Less Water: Is It Possible?" (article sponsored by Almond Board of California) KQED, October 17, 2018, https://www.kqed.org/science/1933028/growing-almonds-with-little-water-is-it-possible.

[g] "Growing Good," Almond Board of California, accessed March 3, 2020, http://www.almonds.com/growing-good.

[h] Adele Peters, "This USDA Lab Is Turning Billions of Pounds of Almond Waste into Beer, Plastic, and Power," Fast Company, March 16, 2018, https://www.fastcompany.com/40544360/this-usda-lab-is-turning-billions-of-pounds-of-almond-waste-into-beer-plastic-and-power.

Downpours lead to more floods, soil erosion, nutrient runoff, and challenges when planting and harvesting crops. Heavy rains can interfere with crop maintenance, increase disease incidence, and affect pest management.[10]

As conditions warm, more precipitation is falling as rain rather than snow in several areas around the globe, including the western US mountain ranges and much of the rest of the country, and scientists expect this will continue.[11] That means some overwintering crops will lose snow's insulating effect. Snow also contains trace amounts of nitrogen, earning it the nickname "the poor man's fertilizer." Snow shifting to rain is a major worldwide concern, with enormous implications for irrigation during the growing season.

With increasing temperatures worldwide, meltwaters from glaciers used for irrigation have increased, benefiting countries such as Peru and Chile, which export more than $5 billion worth of agricultural products to the United States each year.[12] While the glaciers are now providing more water, they will likely be gone in coming decades, affecting agriculture, electric power, and cities.[13] The situation in the Himalaya Mountains, which hold more ice than anywhere else except the North and South Poles, is similar to elsewhere around the globe but on a much larger scale. Changes in glacial meltwaters and precipitation patterns will affect the millions of people who depend on the food grown in the mountains and the vast regions downstream.[14]

In the coming decades, scientists expect hail to increase in size, resulting in more crop destruction in some areas. Hail falling on apples, for example, can render them unmarketable. In the United States, the greatest potential for damage in the South will be in the spring, and in the North in the summer. In contrast, the East and Southeast will see a big decrease in hail damage.[15] In the Netherlands, agricultural hailstorm damage is projected to increase 25%–50% annually and greenhouse production 200% by 2050.[16]

One of the more serious consequences of precipitation changes is contamination of crops when a river or stream overflows its banks and flood fields of vegetables or other flood-susceptible crops. The risk to food safety is high because such flood waters may be contaminated with runoff from upstream cities, farms, and sewage plants. Typically, produce from fields flooded like this must be destroyed, a loss to the farmer and consumer.[17]

Many freshwater lakes around the globe, including the Great Lakes, are now experiencing harmful algal blooms. Warmer temperatures, in combination with nutrient runoff from agriculture and other sources, can promote rapid and large-scale increases in algae. Some of these blooms can pose health risks, interfere with recreation, and constrain the use of such waters for irrigation or drinking.[18] In addition, scientists predict that the Great Lakes and others will become more acidic, but the potential consequences are not well understood.[19]

The Air

The air in which plants grow is changing. Carbon dioxide levels in the atmosphere have increased well over 40% since the beginning of the twentieth century. In theory, excess carbon dioxide should be a boon to agriculture because about 95% of all plants benefit from the "fertilization" effect by increasing their rate of photosynthesis and using water more efficiently. Scientists have demonstrated that under controlled experimental conditions, yields of wheat, soybeans, and rice increased 12%–15% when the carbon dioxide concentration was raised to 550 parts per million, compared with 370 parts per million. However, there are complications. Crop variety, temperature, and drought can all limit the benefits of increasing carbon dioxide. The consensus among experts is that the new extremes in weather will offset most benefits from higher carbon dioxide.[20]

Weeds, like other plants, also benefit from the increase, in some instances becoming more competitive with crops.[21] Ostensibly, weed competition could be negated with weed killers (herbicides), but the evidence shows that increasing carbon dioxide levels may reduce the effect of these chemicals.[22] Likewise, plant-feeding insects might eat more when plants are grown under higher levels of carbon dioxide because nutritional value can be lower and sugar content higher.[23]

Of increasing concern is the predicted decline in the nutritional quality of plants used for human food. When many staple crops, such as wheat, rice, and peas, are grown under conditions that simulate the carbon dioxide levels projected for 2050, they contain less protein, zinc, and iron.[24] Studies focused on rice also showed decreases in B vitamins of 17%–30%.[25] Such decreases would have profound implications, given the millions of people already on marginal diets.[26]

While additional data are needed, nutritional deficits may affect other aspects of the food chain. One intriguing study assessed the protein content of pollen from museum specimens of goldenrod. The research showed a clear decline since 1849, with about one-third of the drop occurring over the past few decades.[27] No one is yet sure how much this protein decrease will affect bees, but these incredibly important players in our food supply are already under stress from loss of habitat and other factors.

Warmer weather puts a wrinkle in prune crops

If you've eaten a Sunsweet prune, Button & Turkovich might have grown it. Tony Turkovich, managing partner at the diversified farm in Winters, California, uses sustainable practices to grow walnuts, oranges, wine grapes, tomatoes, vegetable crops for seed, and field crops. About 10% of the 5,000 ac. (2,000 ha) farm is certified organic, and the award-winning tillage systems developed there reduce emissions, conserve the soil, and save time, labor, and fuel.

The farm grows French prunes—more oblong and fibrous than a plum and very sweet. "Over the last twenty years," says Turkovich, "every so often we've had a warm spring. The prunes bloom in March, and if the weather is in the upper seventies and into the eighties, the pollen doesn't live long enough to pollinate the flowers." He ends up with a very small crop.

According to Turkovich's reading of local data, winters have also been getting warmer over the past fifty years. That presents another problem: not enough "chill hours," which occur when temperatures are 32°–45°F (0°–7°C). In that temperature range, a hormone signals the

The Temperature

It's getting warmer, but the changes aren't simple. Overall, today's cooler regions of the planet are warming faster than the warm regions, and cooler times of day are warming faster than warmer times of the day. In cool regions, cold seasons are warming faster than warm seasons. For example, in the United States, winter temperatures have increased twice as fast as summer temperatures since about the turn of the twentieth century.[28] Also, higher elevations seem to be warming faster than lower elevations.[29]

Globally, nighttime temperatures have increased at a rate about 20% higher than daytime temperatures since 1900, and this has significant implications for our food supply.[30] Yields for rice, for example, decline with trees to go dormant. After 300 to 800 chill hours, depending on the type of tree and the cultivar, that hormone breaks down. Ideally, the weather warms up around the same time, and the tree's buds develop into leaves or flowers.

"When there's not enough dormancy, the bloom will be spread out and be staggered, whereas if there's plenty of dormancy, the tree will come out all at once with a good solid bloom," says Turkovich. "We've been struggling recently to get the chilling hours. It's something that most industry people are concerned about." Lack of chilling hours is already affecting his prune crop, and now he's concerned about his walnuts and grapes.

Solutions include developing prune varieties more tolerant to short chill periods and stepping up nutrition; stronger trees might set enough fruit. Otherwise, Turkovich says, growers will have to grow prunes elsewhere or replace them with a completely different crop that's adapted to the new climate. If prunes cost more to produce, consumers will end up having to pay more for them. "Can we adapt and can we develop the technology soon enough to make a change?" Turkovich wonders. "We're expecting we can."

higher nighttime temperatures, especially during the reproductive phase of the plant.[31] In corn studies, nighttime temperatures raised to what is likely to occur by the end of the century disrupted the pollination process, reducing grain yields by 84%–100%.[32]

With more heat, crops grow faster, and the consequences are not trivial. For example, in California, which grows more than 90% of US and 30% of the world's processing tomatoes, the crop is expected to reach maturity two to three weeks earlier by 2080.[33] This change has implications for the varieties farmers choose, how they manage water, and how they plant and harvest over 247,000 ac. (100,000 ha).[34]

Longer growing seasons may also offer opportunities for new crop lines, or different crops, to maximize production. However, not only are average temperatures increasing but so are the extremes. Very hot weather—heat waves—can harm crops, especially during pollination, which is essential for fruit and seed set. Heat waves are expected to intensify, happen more often, and last longer in the coming decades.[35] High temperature stress is predicted to result in global yield losses of 45%, 52%, and 25% in corn, spring wheat, and soybean, respectively, late in the century.[36]

Because winters are warmer, the US Department of Agriculture (USDA) has shifted its plant hardiness zones up to 100 mi. (161 km) north across the United States. As winters warm, an additional 200 mi. (322 km) shift north is expected in the next few decades. Wheat production will shift north, and as this happens, the southern edge is predicted to get drier and hotter, requiring new and more heat-tolerant strains. Increasing winter temperatures also mean northerly shifts in natural vegetation, as well as new weeds and insect pests.[37]

Warmer winters bring another ominous threat. Fruit and nut trees require a certain number of hours cold enough to keep the tree in its winter dormant phase. Without adequate winter chill, trees set fewer or no blossoms, and yields can drop dramatically.[38]

In 2017, Georgia, the Peach State, had a winter that was simply too warm, resulting in an 85% loss of its peach crop.[39] A warm winter decimated California's pistachio yields in 2015.[40] The winter chill requirements for most fruit and nut crops are known, making it possible to predict when a future winter will be insufficiently cold. Overall, warming winters do not bode well for California, and researchers predict conditions there will become marginal for many of these valuable crops by the end of the century.[41]

Finally, another impact of warming winters is the so-called false spring. When late winter is warmer than normal, buds begin to develop and blossom. Then winter returns, damaging or freezing the buds and flowers. These false springs have caused major crop losses in recent years, though future projections suggest their incidence will decline.[42]

The Soil

"The nation that destroys its soil destroys itself," Franklin D. Roosevelt wrote in 1937 following catastrophic dust storms and flooding in the United States.[43] He was right. Soil deserves respect. A medium for plant growth, a source of nutrients, a habitat for organisms, and a water purifier

and reservoir. It also plays a huge role in modifying the earth's atmosphere. The soil—the skin on the planet's surface—is essential for life on earth.

A healthy soil functions as a vital living system that sustains plants, animals, and humans.[44] It is the most biologically diverse part of the planet, with about four billion bacteria in a handful of fertile soil, many yet to be identified. Simply put, a healthy soil is alive with a multitude of fungi, bacteria, nematodes, insects, spiders, and other organisms that interact with each other in ways critical to the health of the soil and the plants that depend on it. Soils are also diverse in their physical make-up. They can be characterized as clay, sandy, loamy, silty, peaty, and chalky based on the size and composition of their particles. These physical differences are important when planting crops, since they affect the soil's ability to hold moisture, its nutrient content, and how quickly it warms up in the spring.

One key characteristic of a healthy soil is the amount of organic matter. When a plant dies, it decays and forms the organic matter called humus. Because this is a slow process, it essentially holds, or sequesters, that carbon-rich matter for a long time. Some soils, such as in rainforests, typically have above 10% organic matter. Poor soils or those overly exploited can have less than 1%. Soils with higher organic matter absorb more

It's not just food

Just as climate change is affecting our food crops, so is it affecting other products and industries that depend on plants in some way—among them toothpaste, perfumes, clothing, pharmaceuticals, cosmetics, spices, flavors, pet food, medicinal herbs, soaps, natural cleaners, and insect repellents. Here are just a few examples:

Cotton. Cotton is the dominant global fiber crop, with 30 million tons (27 million metric tons) produced in 2018.[a] And cottonseed oil, primarily used in foods, was valued at $3.8 billion globally in 2018, with a predicted value of $5 billion in 2028.[b] Excessively high temperatures, in particular, are predicted to reduce cotton yields by more than 60% by the end of the century.[c] To help address this increasing risk, growers can adopt climate-smart practices and improve water-use efficiency and plant breeders can continue to develop more resilient varieties.[d]

Coconuts. Seventy percent of coconuts, which provide oil for a variety of personal care products, grow in coastal zones that are threatened by rising seas. During the eighteen months a coconut takes to mature, it can be

moisture when it rains and hold that moisture better between rains, which may be beneficial as droughts increase.

The soil also acts as an anchor for roots. Walnut trees, for example, have roots as deep as 7 ft. (2.1 m), and some lettuces 2 ft. (0.6 m).[45] One of the record holders in the United States is a species of juniper with roots about 200 ft. (60 m) deep.[46] The deeper the root system, the greater the access to moisture and the less need for irrigation.

exposed to increasingly severe or prolonged weather events. Scientists studying coconut plants under controlled conditions have demonstrated that they benefit from higher levels of carbon dioxide, but higher temperatures negatively affect them.[e] Increasing dry spells and even greater cloudiness can reduce yield and quality in regions such as Sri Lanka.[f] Irrigation, fertilization, and use of improved varieties can minimize some impacts climate change is having on the world's 32 million acres (12 million hectares) of coconut trees.[g]

Lavender, jasmine, and rose. More extreme weather, including droughts, is challenging these and other sources of fragrance. In May 2018, several lavender-producing families, mainly in Europe, sued the European Union for its role in increasing greenhouse gas emissions.[h] Meanwhile, Grasse, France—where farmers grow the jasmine and rose essential to Dior perfumes—is listed as an area of "extremely high risk" because of climate change.[i]

Medicinal plants. Revenues are expected to reach $129 billion globally by 2023, with an annual growth rate of about 6%,[j] and the changing climate is affecting both medicinal plant effectiveness and their taste. Many species grow in mountainous regions, and warming temperatures mean plants must move up in altitude to remain viable. Some species may not be able to adapt and will become endangered.[k]

(Continued on next page)

The relationship between a plant's roots and the soil is amazingly complex and biologically active. Microbes that live in and on the roots help keep the plant healthy, much like the microbes in the human gut help keep us healthy. Similar types of plants have similar types of microbes; they partner up. Plants can even recruit certain bacteria to their roots to increase their ability to withstand drought conditions.[47] The life under our feet is incredibly diverse and complicated.

To meet the demand for human and animal food, huge swaths of forest are being cleared—more than 502,000 square miles (1.3 million square

(*Continued from previous page*)

Across the globe, large, important industries depend on these and many other plant-based products, meaning climate change will affect more than what's on our plate.

---

[a] "Global Cotton Production Volume from 1990 to 2018," Statista, released August 2018, https://www.statista.com/statistics/259392/cotton-production-worldwide-since-1990/.

[b] "Market Value of Cottonseed Oil Worldwide in 2018 and 2028," Statista, released December 2018, https://www.statista.com/statistics/999312/cottonseed-oil-market-value -worldwide/.

[c] Wolfram Schlenker and Michael J. Roberts, "Nonlinear Temperature Effects Indicate Severe Damages to U.S. Crop Yields under Climate Change," *Proceedings of the National Academy of Sciences* 106, no. 37 (September 15, 2009):15595, https://doi.org/10.1073 /pnas.0906865106.

[d] "Cotton and Climate Change: Impacts and Options to Mitigate and Adapt," International Trade Center, accessed January 23, 2019, 27, http://www.intracen.org/Cotton -and-Climate-Change-Impacts-and-options-to-mitigate-and-adapt/.

[e] John Sunoj et al., "Impact of Climate Change on Plantation Crops: Coconuts," in *Impact of Climate Change on Plantation Crops*, ed. K. B. Hebbar et al. (New Delhi: Astral International, 2017), 22.

[f] Francesco Fiondella, "Climate and Coconuts," International Research Institute for Climate and Society, March 30, 2009, https://iri.columbia.edu/news/climate-and-coconuts/.

[g] "Harvested Area of Coconuts Worldwide from 2010 to 2017," Statista, released January 2019, https://www.statista.com/statistics/1040517/harvested-area-of-coconuts -worldwide/.

[h] Cécile Barbière, "Lavender Farmer Explains Legal Case against EU Climate Policy," Euractiv, July 18, 2018, https://www.euractiv.com/section/climate-environment/news/laven der-farmer-explains-legal-case-against-eu-climate-policy/.

[i] Anne Quito, "The Top Luxury Company in the World Is Fighting to Save the Flowers That Go into Its Perfume," Quartz, retrieved April 30, 2019, https://qz.com/se/perfect -company-2/1172275/the-top-luxury-company-in-the-world-is-fighting-to-save-the-flowers -that-go-into-its-perfume/.

[j] Market Research Future, "Herbal Medicine Market Value to Surpass USD 129 Billion Revenue Mark by 2023 at 5.88% CAGR, Predicts Market Research Future," GlobeNewswire, April 3, 2019, https://www.globenewswire.com/news-release/2019/04/03/1796359/0/en /Herbal-Medicine-Market-Value-to-Surpass-USD-129-Billion-Revenue-Mark-by-2023-at-5 -88-CAGR-Predicts-Market-Research-Future.html.

[k] Manish Das, Vanita Jain, and Suresh Malhotra, "Impact of Climate Change on Medicinal and Aromatic Plants: Review," *Indian Journal of Agricultural Sciences* 86 (November 1, 2016):1375–82, https://www.researchgate.net/publication/309782939_Impact_of _climate_change_on_Medicinal_and_aromatic_plants_Review.

The basis of life is also changing in the oceans

The plants that live in the oceans are as essential to life on Earth as those that grow on land. Phytoplankton, derived from the Greek words *phyto* (plant) and *plankton* (to wander or drift), are at the base of the ocean's food chain. Like plants on land, they use the sun's energy for photosynthesis—but they float in the oceans. They also produce over half the oxygen on which the planet's animal life depends, and they absorb carbon dioxide from the atmosphere.[a] The warming oceans are becoming more stratified, or layered, and this prevents the upwelling of nutrients the phytoplankton need from deep in the ocean. In some areas, phytoplankton have declined 20%.[b] And scientists are concerned about what lies ahead for these critically important creatures. A 5°F (3°C) increase in temperature by 2100 will likely cause global changes in the make-up of phytoplankton. These shifts will have serious implications for the ocean's food web.[c] When researchers subjected green algae to the worst conditions expected in ocean waters by the end of the century, the algae photosynthesized less efficiently.[d] These impacts on phytoplankton and algae suggest that fundamental changes may be lurking in the ocean food chain that would affect tiny phytoplankton, the small creatures that feed on them, the fishes that come next, and all the way to the top—*Homo sapiens.*

---

[a] "How Has the Ocean Made Life on Land Possible?," National Oceanographic and Atmospheric Administration Office of Exploration and Research, accessed September 8, 2019, https://oceanexplorer.noaa.gov/facts/oceanproduction.html.

[b] Mathew Koll Roxy et al., "A Reduction in Marine Primary Productivity Driven by Rapid Warming over the Tropical Indian Ocean," abstract, *Geophysical Research Letters* 43, no. 2 (2016):826, https://doi.org/10.1002/2015GL066979.

[c] Stephanie Dutkiewicz et al., "Ocean Colour Signature of Climate Change," *Nature Communications* 10, no. 1 (February 4, 2019):1–13, https://doi.org/10.1038/s41467-019-08457-x.

[d] A. Gomiero et al., "Biological Responses of Two Marine Organisms of Ecological Relevance to On-Going Ocean Acidification and Global Warming," abstract, *Environmental Pollution* 236 (May 1, 2018):60, https://doi.org/10.1016/j.envpol.2018.01.063.

kilometers) between 1990 and 2016.[48] When that happens, some of the carbon dioxide released comes from the decomposition of decaying plant material on top of the soil. The organic matter in the soil also decomposes, and since there is no more input of vegetation from trees above, the result is a rapid decline in soil carbon. Similarly, the first cultivation of previously undisturbed land—and land that is repeatedly cultivated without a return of organic matter—results in a loss of carbon from the soil.

Soil and climate change are linked because 80% of all the carbon in terrestrial systems is found in the soil, primarily in organic matter.[49] An estimated 1.4 billion tons (1.3 billion metric tons) of carbon are released into the atmosphere each year from changes in land use, such as the cultivation of land for agriculture and the clearing of forests.[50] Experts agree that soil is a major source of carbon dioxide but also has potential to sequester it at a scale that could help address climate change.[51]

The Animals We Eat Also Need Plants

We eat crops fresh from the farm, such as asparagus and sweet corn, as well as crops that have been processed, cooked, or fermented, creating some of our favorite foods and beverages. At the same time, billions of hogs, beef cattle, sheep, and fowl also depend on crops. In 2018, the number of pigs worldwide was estimated to be 770 million, and cattle about 1 billion.[52] To sustain these huge populations requires a lot of land. Worldwide, 77% of farmland is used for animal feed production and grazing.[53] With rising incomes in developing countries, demand for animal products is expected to increase. In the United States, domestic animals consumed 236 million tons (214 million metric tons) of animal feed in 2016. Feed for cattle, broilers, and pigs led the way.[54] Increasing temperatures and carbon dioxide levels and changes in precipitation patterns don't just affect the crops we humans eat—they also affect the food supply for an enormous population of domestic animals.[55]

Pollinators

To reproduce, many plants need pollinators such as bees and other insects. When a bee visits a flower to obtain nectar and pollen, some pollen from the flower's male part, the stamen, sticks to the bee's body. When it visits another flower, some of that pollen rubs off onto the flower's female organ, or pistil. The result is fertilization and seed development. Several types of bees are important pollinators, as are many species of flies, wasps, butterflies, beetles, and moths. Insects' role as pollinators cannot be overstated. Worldwide, about 75% of crop plants and 94% of wild flowering plants benefit from their services.[56] The list of crop plants that require or benefit from pollination by bees and other insects is long and includes almonds, beets, plums, currants, pumpkins, grapes, and sunflowers.

A lot is changing for pollinators as the climate changes. The wild and crop plants that pollinators need might bloom earlier because of warmer conditions, but the pollinators might not be synchronizing their life cycles with the plants' new bloom time or new location. Consequently, some pollinators could decline or even go extinct. Experts agree that pollinator populations are in general decline in many regions of the world due to a variety of factors, including loss of wild habitats, invasive species, some pesticides, the spread of diseases and pests, and climate change.[57] While some flexible pollinator species may persist and thrive, the new conditions will stress many species so critical to our food supply.

Pests

As climate change unfolds, the stress on crops from insects, weeds, and plant diseases will likely intensify.[58] Pests already cause extensive losses to agriculture, with weeds alone responsible for an estimated 34% loss globally.[59] Higher temperatures, changes in precipitation, and, in some cases, increasing levels of carbon dioxide will affect the distribution, frequency, and severity of pest outbreaks, as well as their management costs.

With warmer growing seasons, insect pests will develop faster and may have more generations per year, potentially resulting in more crop damage. Warmer winters will allow more insect pests to overwinter and expand their ranges north into new territory. Some species are now appearing much earlier in the season than in the past. Studies in the United Kingdom determined that the first flight of aphids (tiny plant-sucking insects) occurred about thirty days earlier in 2012 than in 1965.[60] Certain pests might emerge sooner in the spring than their natural enemies, or vice versa, meaning the pests would simply escape the suppression that the natural enemies historically provided. Scientists realize that these complex interactions

Insect likes these new, warm digs

Tom Turini tromps around large farms in Fresno County looking for trouble, and sometimes he finds it. As a University of California vegetable crops advisor, he offers advice to growers who are raising melons, tomatoes, and other commercial crops. So imagine his concern when they began seeing huge numbers of a tiny pest they'd never worried about before.

The silverleaf whitefly, a yellow insect with white wings, feeds on plant sap. It stunts or kills plants while exuding a sticky honeydew that invites black sooty mold. For decades this whitefly has been a major problem in the low-desert regions of Southern California, where temperatures stay above freezing and the pest can thrive year-round. In Central California where Turini works, however, sustained winter freezing has killed silverleaf whiteflies in the past.

Not so now. Turini has been measuring the accumulation of heat that determines how fast insects and crops will develop. "From 2014 to

may result in some insect pests becoming a greater crop threat and others a lesser threat.[61] The incidence of crop diseases will likely also increase because of warmer winters, earlier spring conditions, and changes in precipitation patterns.

The Food Supply Chain: Beyond the Farm

Climate change will affect almost every stop along the food supply chain from the farm to our dinner table. Like so many other components of the climate change story, various parts of the supply chain are facing new risks and becoming more vulnerable.

Trains, trucks, ships, and planes transport massive quantities of food every day, and they depend on a vast infrastructure of roads, canals, railways, runways, bridges, canals, and shipping terminals. The transportation sector is also growing, driven by increasing population and expansion of urban

2018," he says, "every year has been higher than the thirty-year average." The whitefly completes its life cycle from egg to young in only twenty-one days, so populations can build up fast when it's warm. Silverleaf whitefly pressure has now reached economically damaging levels.

In 2018 a grower with a 160 ac. (65 ha) field of cantaloupes could not fully control the whiteflies. Turini describes that scene unfolding: "You don't even see the bottom of the leaf; it's just solid with whitefly nymphs. You walk up to these stunted plants that are desiccated, essentially, and you can see the honeydew darkening the ground. It's massive."

Twenty acres of that grower's cantaloupes died or produced no fruit. "They lost quality and yields in many other fields, too," says Turini.

Silverleaf whiteflies feed on many crops, including honeydew melons, broccoli, cabbage, cotton, and tomatoes, so growers must consider what they grow next to their melon fields, lest additional crops become infected.

"It's a different way of thinking that needs to be one of the motivators for where you put your crops," says Turini. "For the coming season, I met with one of the growers that has thousands of acres and talked to him about the importance of planning. You have to avoid sources of white-flies. So he's going to do it. He said, 'We can't lose again the way we lost last year.'"

centers worldwide. Storm surges, floods, and extreme weather are already affecting nearly every form of transportation and related infrastructure, much of which was not designed in anticipation of a changing climate.

The gradual rise in sea levels and an increasing incidence of storm surges associated with hurricanes pose direct threats to transportation infrastructure in low-lying coastal regions around the world. At risk are ports and airport facilities, roads, rail lines, tunnels, and underground transit systems. In the United States over 60,000 mi. (97,000 km) of roads are exposed to coastal storm surges.[62] Vehicle delays will likely increase over tenfold by 2060 along the US East Coast, with obvious implications for transport of food, especially perishable food.[63]

The increasing incidence of extreme weather across the United States interferes with most modes of transportation, from road traffic to airline flights. Droughts, which are expected to increase in coming decades, do the same; a 2012 drought lowered Mississippi River water levels, greatly

The food system contributes to climate change

Agriculture and the global food system are not only affected by climate change but are also major contributors to it. The entire food production, distribution, packaging, storage, and retail system, including the manufacturing and distribution of seed and fertilizers, contributes about 25% of the total human-caused greenhouse gas emissions. Of these, most comes from agricultural production.[a] Cattle produce greenhouse gases, as do rice paddies, deforestation, burning biomass, poor manure management, and the use of synthetic fertilizers.

[a] J. Poore and T. Nemecek, "Reducing Food's Environmental Impacts through Producers and Consumers," *Science* 360, no. 6392 (June 1, 2018):987, https://doi.org/10.1126/science.aaq0216.

encumbering barge traffic. High waters and flooding also interfere with transportation on this major commercial artery.[64]

Worldwide, chokepoints—critical junctures on transport routes with exceptional volume—are at increasing risk. These include canals, straits, and inland transport infrastructure essential to the passage of crops, fertilizer, and other products. If climate change events or associated social or political upheaval causes interruptions at one or more global chokepoints, price spikes, shortfalls in supplies, and spoilage of commodities could result. For example, the Strait of Malacca is one of the most important shipping lanes in the world, and its disruption would affect the transport of one-fourth of the global soybean exports destined for animal feed in China and Southeast Asia.[65]

Increasing temperatures and more extreme weather will directly and indirectly affect food processing and storage facilities, as well as labor. For example, rising sea levels or flooding could shut down electric generation plants or electricity transportation systems, disrupting food storage and processing. Also, studies in India showed that every increase of about 2°F (1°C) can increase worker absenteeism and decrease labor productivity by 2% to 4%.[66] Higher temperatures will also require increasing investments in air conditioning and cooling for products, employees, and customers. All this will likely make our grocery bills increase, as the tolls that extreme weather and increasing temperatures take all along the food chain add costs over time.

Summing It Up

Climate change is impairing the entire food system, from a plant's needs—air, water, soil, and the right temperature—to how we transport food and food products around the globe. It has already affected food production globally and will intensify in coming decades.[67] Climate change and other factors are making more vulnerable a system that currently relies on just twelve plant and five animal species to provide 75% of the world's food.[68] Globally and in the United States, frequent droughts, rising temperatures, changes in precipitation patterns, and shifts in the distribution of pests will threaten the foods we need and love.

Small incremental actions and large-scale ones can help us solve this grand challenge. We must reduce greenhouse gas emissions at a scale and in a time frame that will make a difference. We can strive to adapt our agricultural and food systems to the new normal and reduce the impact these systems have on the climate.

Now we focus on the heart of this book: how climate change is affecting the many items on the menu and what's being done to help keep them there. We'll start with distilled beverages and end with dessert and coffee.

THE MENU

We now look closely at how the menu is changing—the heart of the book. As if describing a meal, we show how drinks, salads, main courses, sides, and dessert and coffee are changing. We highlight key ingredients along the way, such as hops for beer, olive oil for salads, beef as a main course, rice as a side, and vanilla for desserts. We celebrate foods and depict how integral they are to our history, our cultures, and our economies. Then we answer questions about what is changing now and in the future, and what needs to be done to keep the menu stocked.

Many people enjoy wine, beer, and distilled spirits. In the United States we each drink 20 gal. (74 L) of beer and about 2 gal. (8 L) of distilled spirits each year, and 10,000 wineries attend to our wine needs. However, grains for beer are faced with excessive heat; changes in water quality and quantity are affecting spirits; and wine grape production is shifting to cooler climes. Salads are not immune, with avocados and olives facing more stressful higher temperatures and water shortages. The main course follows, with emphasis on beef, poultry, and fish as well as side dishes and how they are facing increased risks from a changing climate.

We finish our imaginary meal with desserts and coffee. Globally, two billion tons of sugarcane are produced annually, the unique chemical properties of which make it perfect for use in desserts. Eight million tons of chocolate—food of the gods—is enjoyed worldwide, along with milk products like ice cream and cheese. The problems? Hot cows give less milk; intensifying storms in Madagascar disrupt production of vanilla; and our beloved coffee is at more risk because of spreading pests.

THE MENU

Despite the many challenges, we share a message of hope and inspiration. We describe how producers, scientists, people in the food businesses, and many others are working hard to find ways to mitigate the impact food production has on the climate and to keep the foods we love and need on the menu.

Beer, Wine, and Spirits

Raise Your Glass

Whether dinner is a barbecue in the backyard, an evening meal with family and friends, or a formal gathering with business associates, there is nothing like an alcoholic beverage to stimulate the conversation and the appetite. As we dive into how climate change is affecting our dinner menu, let's start our imaginary meal with a toast. Alcohol has been part of human culture for thousands of years, typically in the form of wine or beer. Throughout history, people across the globe have enjoyed fermented beverages in taverns and banquets, at celebrations, and after a day of hard work. Sharing a drink has strengthened friendships, new and old, cemented enlightened thinkers' grand ideas, and made days that much sweeter.

Archeologists date the earliest known alcoholic beverage to the Yellow River Valley of China nine thousand years ago.[1] Around the same time, denizens of the Middle East started making wine and beer. Spirits were frequently offered as a health tonic, even as competing voices counseled moderation.

Beer

A quart of ale is a dish for a king.

SHAKESPEARE, *THE WINTER'S TALE*

Ancient pottery jars show evidence of beer from seven thousand years ago. All it took to get fermentation started was grain, the right kind of sugars, and wild yeast dropping in from the air. And it appears that the ancients prioritized brewing, as they seem to have domesticated grains to make beer rather than to supply food.[2]

Some early beers were so thick and gruel-like that anyone who drank them needed a drinking straw to suck up the liquid and leave the solids behind. Later, hops added flavor and helped preserve the beer. Today, people brew beer at highly sophisticated industrial-scale facilities, at thousands of smaller regional breweries, and at home, some using very old methods. The choices for the beer consumer are nearly unlimited. What type will it be—ale, lager, stout, or porter? With a full body, hoppy finish, or fruity, nut, or toffee flavor? Styles include amber, brown, cream, golden, honey, red, and wheat. No one really knows how many kinds of beers there are in the world.

How Is Beer Made?

Beer is very easy to drink but not to make, especially at a commercial scale. It's energy intensive, requiring considerable heating and cooling along the way, and it is also quite water intensive, sometimes requiring ten parts of water to make one part of beer. During malting, grain is heated, dried, and cracked. This process is important to the flavor and style of beer. The grains then go through mashing, where they are steeped in hot water, and sugars are released. Once the water is drained, the sweet liquid remaining, called wort, is boiled. Hops and sometimes herbs and spices are added. The wort is cooled, filtered, and put in a fermentation vessel, to which the yeast is added. The yeast turns the sugars into alcohol and carbon dioxide. Last, the beer is bottled and aged for a few weeks to a few months, and then it's ready to drink.

Beer can be made with barley, corn, rice, and wheat, and sometimes even sorghum, cassava root, potato, and agave. But barley is the most widely used beer grain because of its high starch content. The oldest cultivated grain, barley is well adapted and grown in climates ranging from sub-Arctic to subtropical. The world's heavy hitters in barley production are the European Union, followed by Russia and Canada.[3] The United States is the seventh-largest producer worldwide. In 2017, barley was grown on almost 2 million acres (809,000 ha) in the United States and valued at $614 million, with Idaho, Montana, and North Dakota leading production. Three-quarters of US-grown barley is used for food and malt purposes, the balance for animal feed.[4]

Hops, the flowers of the hop bine, give beer its bitterness and flavor. The hop bine is in the same plant family as marijuana, but that is a different

story. Hops have been grown in the United States since the 1600s. Today the Pacific Northwest, with its fertile soils, favorable climate, and adequate water, is home to almost all domestic hop production, but many small operations exist across the country. In 2019, Washington, Oregon, and Idaho combined grew hops on 56,500 ac. (22,865 ha). That year, total US production was valued at nearly $637 million.[5]

Water makes up 95% of beer and can also affect its flavor. Soft or hard water and water containing unique flavor-affecting minerals can give beer a regional character. For example, the English town of Burton on Trent is known for its exceptionally good bitter beers, which are attributed to the unique sulfate-to-chloride ratios in the water. To control flavor, large breweries might remove all minerals from the water but then add them back to their own specifications.

Who Cares?

The world's beer drinkers consumed an estimated 49 billion gallons (186 billion liters) in 2017—equivalent to almost 300 billion bottles. Asia represents the largest slice of the world's market, about one-third, while the Czech Republic holds the record for annual consumption per person, about 48 gal. (183 L). The US share of the global market is about 13%, and per capita consumption is 20 gal. (74 L) per person.[6] The beer industry is a big business.

What Is Changing for Beer?

In 2017, Montana farmers experienced excessive heat and a lack of rain, and those who were growing rain-dependent (not irrigated) barley took a hit. Their barley had small, underweight grains—good for animal feed, not beer, and worth half as much. A couple of years earlier, an unusual rain came near harvest time, and the mature barley sprouted in the field instead of in the malthouse, causing extensive losses. Such crop-stressful conditions will worsen with climate change. For example, Montana temperatures are expected to rise 4°F–5°F (2°C–3°C) by midcentury.[7] A 2018 report indicates that worldwide barley yields will decline 3% to 17% by the end of the century, resulting in less beer and higher beer prices, depending on location.[8]

Washington State, which leads US hop production, has also experienced increasing temperatures and drought.[9] Mountain snowpack, a critical source of irrigation water during the hop-growing season, has already declined dramatically and is expected to continue to decline as temperatures rise.[10] Access to this water source could become contentious in the future. For example, between 1979 and 1999, there were three water-short years in Washington's Yakima basin, resulting in reduced water access for those with junior (lesser) water rights. Toward the end of the present century, restricted water delivery is projected to occur in as many as fourteen years out of twenty.[11]

Water shortages can also harm brewers themselves, who might draw water from rivers, reservoirs, and other surface sources. After multiple years of drought in Northern California, breweries were anxious about having to use mineral-heavy groundwater, which taints beer with an astringent taste.[12]

Droughts are also affecting hop yields in Europe, where in 2015 German production was down over 25% from the previous year.[13] The United States and Germany each produce about one-third of the world's crop—so that reduction was a big deal.

In some cases, a warming climate can directly affect the brewing process. In 2015, a leading Belgian artisan brewer halted production because of a very warm fall. Typically, he would ferment Lambic beers in the open to cool them and infuse them with wild yeasts drifting in the air, but warmer fall and winter conditions are shortening his brewing season. Ideally, brewers need five cool months to brew this beer. Losing a week or so is tolerable, but several weeks make production much more challenging.[14] Unfortunately, all seasons will be warming in Belgium, with winters expected to rise as much as 8°F (4.4°C) by the end of the century.[15]

What Is Being Done to Keep Beer Mugs Full?

To overcome some of the increasing climate threats to barley, researchers and others are developing new varieties that are more drought- and heat-tolerant, resistant to disease, and less likely to sprout prematurely. Overall, growers prefer resilient types that can handle multiple stressors, such as a new line developed by crossing wild types with a standard variety. The cross handles salt, heat, and drought.[16] Surprisingly in a warming world, one of the more promising traits is cold tolerance, allowing winter-grown barley to be planted in the fall and harvested early in the summer of the next year, before hot temperatures arrive. Some winter barley is already being grown in Oregon and other places with relatively mild winters.[17]

Hop hunters are also on the prowl for new, hardier varieties that can handle warming temperatures. Unfortunately, the types of hop most at risk from extreme temperatures and drought are those responsible for some of the hoppy flavors popular with beer drinkers. More frequent hailstorms are also damaging the delicate hop cones. Increased resistance to some of these new challenges might be discovered in the mountains of Arizona, where wild hops are common.[18]

Not all beer industry solutions are technical or economic. Brewdog, an independent brewery in Scotland, sent a lighthearted political message following President Trump's withdrawal of the United States from the Paris Agreement. They created Make Earth Great Again, a beer that incorporates ingredients climate change has affected: water from melting Arctic ice caps and flavoring derived from endangered Arctic cloudberries. Fermenting this beer at a high temperature reflects increasing global temperatures. Brewdog donates all proceeds to a UK-based climate change charity that works with communities to take local action.[19] Similarly, New York's Sixpoint Brewery encourages consumers to reduce their carbon footprint; they note on their Global Warmer beer can, "Extended refrigeration at retail magnifies beer's carbon footprint. Please enjoy as soon as possible."

Finally, several major breweries have signed onto a climate declaration to raise awareness of the new risks facing their industry.[20] Brewers are also assessing greenhouse gas emissions, switching to renewable energy sources, and conserving water.[21] In Australia, water is now considered a precious resource, and a major brewer has launched a beer made from cassava because it is less water dependent than other malting crops. Some companies, for example the Craft Brew Alliance, have reduced the amount of water needed to make a gallon of beer by over half.[22]

What Does the Future Hold?

As we move into uncharted territory with increasing temperatures and more storms, droughts, and floods, this ancient and refreshing beverage won't be the same. It will no doubt still be on the menu, but where it comes

Hunting hops to protect them

In the Sky Islands—a teardrop-shaped region where Arizona, New Mexico, and Mexico touch—Taylan Morcol moves among rocky fields looking for wild hops. The mountains here rise above the desert like islands in an ocean. Plants and animals thrive on these cool, moist slopes but cannot survive in the dry heat below. They are separated from relatives on neighboring mountains and evolve distinct characteristics, like Darwin's finches in the Galápagos.

Morcol and other scientists are cataloging Sky Island hop plants, determining whether they differ from one another genetically and chemically. He and his colleagues hope to eventually breed varieties that resist diseases and withstand drought, heat, and other climate-related stresses. The area's wild hops may have very high genetic diversity, but climate change and human encroachment threaten the habitat. "There's a pressing need to characterize those hops while they're still there," he says. So he hunts hops.

Before venturing out, Morcol consults herbarium records, researchers' field notes, and iNaturalist—an online library of photographs and observations from 280,000 citizen scientists. He and his colleagues focus on one variety: *Humulus lupulus* var. *neomexicanus*. Like other hops, *neomexicanus* has deep roots that tap into water sources in rocky ravines and high-altitude talus where little else will survive. It is selective about where it will grow, but the variety seems to be hardy, even weedy. "Plants can be cut back to nothing and they'll grow back the next year," says Morcol.

When he discovers a new cluster of *neomexicanus*, he takes samples of the seeds, leaves, and rhizomes (the underground stems). Then he grows the rhizomes into plants in a greenhouse at the City University of New York Graduate Center, where he is a doctoral student. Hopsteiner, a global hops breeder and supplier, funds Morcol's field team, and the US Department of Agriculture's National Clonal Germplasm Repository will store the seeds the project produces and make them available to breeders and researchers worldwide for free.

"Resiliency of agricultural systems depends in part on how much genetic diversity there is," says Morcol. "The more diversity, the greater chance that things will survive if something catastrophic happens or the environment changes. That's true of many crops, including hops." He hopes to preserve the diverse genetic material of wild hops trying to survive in the threatened Sky Island habitat. "If it's lost, then it's lost," he says. "It's gone."

from, who brews it, and what styles remain will have changed, thanks to our changing climate.

Wine

Wine is bottled poetry.

ROBERT LOUIS STEVENSON, *THE SILVERADO SQUATTERS*

Many restaurant wine lists are long and varied, and selecting the right wine can be challenging. Though wines may be made from different fruits, most are grape based. Wines develop their distinct tastes and aromas from the type of wine grape, where it's grown, the local climate, the soil type, how the wine is fermented, and whether and how it's aged. The main characteristics of wine are sweetness, acidity, alcohol, tannin, and body, but dozens of other descriptors include thin, hollow, elegant, earthy, chewy, voluptuous, zippy, stalky, dill, buttery, and green. No wonder wine satisfies the palates of

millions—some combination of these characteristics appeals to almost every one of us.

Wine has been around for about eight thousand years, analyses of compounds absorbed in pottery in the South Caucasus region show.[23] From there it spread across much of Europe, and Spanish missionaries introduced it to the Americas in the 1700s. Those looking for gold in California in the mid-1800s brought grapevines with them from the US East Coast. Today, wine grapes are grown widely around the globe.

Wine making has come a long way over the centuries, but the basic steps remain the same: harvest and crush grapes, ferment them into wine, possibly age it, then put it in a bottle. Although the process is simple, innumerable twists and turns can take place. Workers harvest grapes by hand or machine during the day or, if daytime temperatures are high, at night. Red grape skins as well as crushed grapes may go into the fermentation vat to give the wine a red color. White grapes may be fermented with or without their skins.

During fermentation, most grape sugar turns into alcohol. Here again the process can take many different directions, but generally winemakers add yeasts to the vats to ensure predictable fermentation. Vintners age wine for months or years in stainless steel tanks or oak barrels. The source of oak and how the barrel is crafted influences the wine's flavor. The wine is bottled, perhaps aged some more, then sold and consumed. In 2018, vintners produced an estimated 7.7 billion gallons (29.2 billion liters) globally.[24]

Who Cares?

Italy, France, and Spain produced the most wine in 2018, followed by the United States, which produced about 10%.[25] Globally, the wine industry is expected to grow in coming years, with increased demand for New World wines from countries such as New Zealand, Chile, the United States, and Australia. The French and Italians compete for the highest annual consumption rate, about 12 gal. (45 L) per capita in each country.[26]

In the United States, ten thousand wineries—some in every state—are the backbone of an industry that employs close to one million people and contributed $220 billion to the economy in 2017.[27] Sales of domestic table and imported wines totaled almost $70 billion in 2018, with California's forty-five hundred wineries providing 86% of the domestic supply.[28]

What Is Changing for Wine?

Wine grapes, especially premium quality wine grapes, are very sensitive to temperature changes. Under warmer conditions, grapes may be ready for harvest earlier and have higher sugar levels, lower acidity levels, and different aromatic compounds. In combination with less-dependable rainfall patterns, higher temperatures will likely lead to lower yields and poorer grape quality in many warm-climate wine regions. Because of increasing dry spells in Europe, growers must irrigate to sustain the vines, a practice historically banned because it compromises *terroir*—the mix of local land, soil, climate, and culture that results in a specific wine from a given locality having a distinct sense of place.

The wine industry is already feeling the effects of climate change. In France, for example, harvests are about ten days early in some regions because warmer temperatures are speeding up crop development. However, some growers don't harvest early but leave the grapes on the vine to

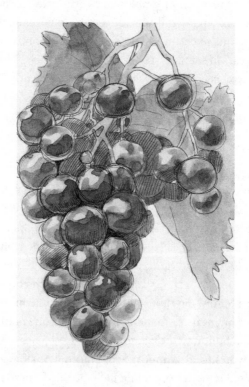

rebalance the fruit chemistry the higher temperatures have disrupted.[29] Overall, for every 1.8°F (1°C) increase in France, the harvest date is about six days earlier.[30] In Europe generally, flooding events have quadrupled since 1980 and extreme heat waves, droughts, and forest fires have doubled, making wine grape production more challenging.[31]

In 2017, world wine production dropped about 8%, mainly owing to extreme weather, especially in Italy, Spain, and France, where production was down 15% to 23%.[32] Some consider 2017 one of the toughest on record for the wine industry.[33] Fires that year took a toll on vineyards and wineries in California, although the impact would have been worse if the harvest had not occurred a little earlier than normal. Grapes on the vines during the fires were tainted and, in some instances, left unharvested. The air quality even after the fires were extinguished was too poor to allow workers to harvest grapes. A few wineries were damaged, with one losing about 15% of its bottled wine.[34] Producers and many others are concerned that the new "fire season" will affect tourism, a business that draws twenty-four million people a year to the region.[35]

Climate change will have enormous impacts on future wine production worldwide. Some scientists suggest that the world's major wine-producing regions may see declines of 25% to 73% by 2050.[36] In the United States, some experts estimate an 80% decline in premium wine grape acreage by the year 2100.[37] Conversely, in recent years, some regions with warmer and drier conditions experienced improved wine quality, although it came with a yield loss.[38]

A much warmer and drier future does not bode well overall. Like many other fruit crops, grapes need a given period of dormancy or winter chill to set fruit. Climatologists predict that by 2100, the winter chill in California will be reduced to five hundred hours, which is about the minimum number of chill hours grapes require.[39]

Warming winter temperatures mean some grape pests are expanding their ranges northward. The glassy-winged sharpshooter is a current threat in California, where it spreads Pierce's disease, causing rapid deterioration of vines.[40] In the future this insect is likely to be a problem in Oregon. More insect and disease pests will surely move north as conditions continue to warm.

Wine producers face other risks too:

- Rising sea levels threaten some wine-making regions in France (Bordeaux), Portugal, and elsewhere.[41]
- Droughts are slowing the growth of cork oaks, the source of millions of corks for wine and other spirits.[42] More droughts are expected in the future.
- Hot weather is hard on laborers, so labor costs—and the price of wine—may rise.[43]

Climate change is also affecting table grapes, which we eat fresh or dried as raisins. In 2019, growers worldwide produced 23 million tons (21 million metric tons) of fresh table grapes, up about 4% despite losses due to heavy rains in the European Union. In the coming years, higher temperatures and water limitations will continue to affect table grapes. The United States produced about 998,000 tn. (905,000 t) in 2019, nearly all in California.[44]

What Is Being Done to Keep the Wine Flowing?

With growing conditions becoming difficult to untenable in the coming decades, some vintners could move northward or seek higher elevations with cooler temperatures. Production may decline in traditional grape-growing regions and expand in other regions. For example, vineyard acreage in northerly western North America, northern Europe, and New Zealand could increase.[45] The northeastern United States will likely contend with more heavy rainfall events and other weather extremes, but it may also see an expanded wine industry, especially for red wines.[46] Scientists predict that cool-season varieties will continue to move to higher latitudes with longer day lengths during the growing season.[47]

Growers in traditional regions might still be able to grow grapes, but new investments in irrigation, shade systems, and misters may drive up costs. Vintners drawing water for irrigation might deplete local and regional resources, so they will need to select vineyard sites that minimize negative impacts. Winemakers less concerned about *terroir* could purchase grapes grown both locally and farther away. Wine grape growers in Australia are experimenting with another option: growing varieties already adapted to warmer and drier conditions from the hotter regions of Europe.[48]

Experts generally agree that developing new varieties more resilient to variable conditions is the best way to adapt to climate change. A large pool of genetic diversity exists: eleven hundred grape varieties are available for commercial production, yet only twelve are widely used.[49] For example, 75% of the grapes currently grown in China are Cabernet Sauvignon because that is what consumers prefer.[50] In changing local climates, plant breeders can instead select for varieties with traits that align with new conditions, say, when vines leaf out or set flowers, or when berries increase in sugar content, change color, or mature. With adequate research, vineyard testing, marketing, and education, traditional wine-grape growing regions might remain viable—as long as wine drinkers are willing to accept new wines from their treasured regions. Breeders who tap the pool of wine grape diversity might also help find the best varieties for new locations as grape production moves up in latitude and altitude.

Vineyard managers are already adapting to the more extreme conditions. The California Sustainable Winegrowing Alliance has produced *A Winegrower's Guide to Navigating Risks*, which addresses drought, heat, heavy rain, flooding, and wildfires, along with ways to reduce greenhouse gas emissions.[51] Among many recommendations, it suggests that growers

- use drip irrigation and water-monitoring tools to apply water when the vine needs it;
- enrich the soil with organic matter to help soak up and hold rainwater;
- use cover crops to help sequester carbon;
- reduce fossil fuels and conserve energy; and
- consider alternative energy sources, such as solar.

Jackson Family Wines in California is reducing fossil fuel use and adapting to the new conditions. The owners irrigate their grape vines efficiently and have installed over one hundred reservoirs to hold water and help them through dry spells. They also fly drones over the vineyards to assess irrigation needs. In their winery they use ultraviolet light instead of water to clean tanks, saving 28 million gallons (95 million liters) of water each year. They are also expanding north into Oregon, where conditions are cooler and more water is available.[52]

Vineyard grows grapes with resilience

On the eastern shore of Seneca Lake, Silver Thread Vineyard stands ready for what climate change might bring: torrential rains, temperature swings, and hotter summers.

In 2011, Shannon Brock and her husband purchased this 8 ac. (3 ha) estate in Lodi, New York, pledging to use sustainable practices to make the best wine possible. As Brock explains, "The wine is supposed to be an expression of the place where it was grown. Anything we're putting on those grapes—or into that wine—that might alter the taste or mute the natural flavor coming through is a detriment." From their Riesling, Chardonnay, Pinot Noir, Merlot, and other grapes, the Brocks produce three thousand cases of whites, reds, and rosés each year.

The Brocks have made several choices that help their grapes thrive in a changing climate. Complete cover cropping means no exposed dirt under their vines, says Brock. "We have native grasses that grow between the rows. Underneath the vines, where the typical vineyard would spray herbicides, we have a low-growing grass." That fescue outpaces weeds and doesn't grow tall enough to compete with the vines, so the Brocks don't use herbicides. In addition, it stores water and nutrients much better than bare soil and protects against erosion.

Noting that intense storms seem more frequent, Brock cites a 2018 tempest when 9 in. (23 cm) of rain fell in twenty-four hours. "There were several vineyards around the region that experienced some pretty severe erosion," she says. "We had virtually zero erosion in our vineyard."

Temperature swings from warm to frigid in a day's time can freeze and split grapevines. "Part of the way we've tried to insulate ourselves from that," Brock says, "is by choosing a vineyard in a prime location that's known for being a very protected spot." The deep lake moderates temperatures at Silver Thread. If someday that doesn't work, Brock will consider planting varieties that are more resilient.

Warmer temperatures might bring new opportunities. "As we experience higher average temperatures and longer growing seasons," Brock says, "I don't think it's an accident that the quality of Finger Lakes red wines has improved. . . . I don't know how much of it is *business* climate or how much of it is the *actual* climate, but in the past few years Americans have preferred reds over whites and are willing to pay more for them." Though red wine grapes take additional labor, the Brocks are looking into hybrids they can grow organically, especially as the climate becomes more favorable to red grapes.

Vintners are changing across the globe. Spain's Familia Torres has committed to reducing emissions per bottle by 30%. Their actions include use of a biomass boiler that burns old vines and pruning canes, reducing natural gas use by 95%, use of recycled water, wide-scale use of hybrids and electric vehicles, and installation of solar arrays.[53] Torres and Jackson wineries founded International Wineries for Climate Action and encouraged other wineries worldwide to help address the challenges posed by a changing climate.[54]

As the Romans said, "In wine there is truth," and the truth of climate change is upon us. In response, the wine industry is shifting toward a more sustainable business model worldwide.[55] Consumers can also help ensure the success of the wineries they love by accepting inevitable changes in their favorite wines.

Spirits

Walk through the local liquor store or look behind the bar at a local pub and you'll see the wide-ranging choices of distilled spirits. You can have a sip of tequila, a dram of Scotch, or a snifter of bourbon or enjoy a Margarita, Manhattan, Bloody Mary, Negroni, Mimosa, Gimlet, Sidecar, or even a Mint Julep, to name but a few.

How Are Distilled Spirits Made?

The making of a distilled drink or liquor starts with the fermentation of grains, fruits, or vegetables. Next comes distillation, which is the separation of alcohol (ethanol) from water. As long as the temperature in the still is above the boiling point of alcohol but below that of water, the alcohol boils off to be cooled and condensed, a process distillers repeat again and again to increase the alcohol content. A spirit has an alcohol content of 20% or more.

Who Cares?

Many countries have a national distilled spirit as part of their culture—for instance, Mexico's tequila, Jamaica's rum, England's gin, Russia's vodka, Germany's schnapps, and the United States' whiskey. The global distilled spirit industry is a significant economic engine, valued at $443 billion in 2019 and expected to grow to over $500 billion by 2023.[56] South Koreans lead annual consumption of spirits with 8.5 gal. (32 L) per capita, while Russians drink about 5 gal. (20 L), per person, and people in the United States about 2 gal. (8 L) each.[57]

Distilled spirits constitute 37% of the US alcohol industry, beer 46%, and wine about 17%.[58] How much spirits people in the United States drink per person each year varies considerably by state: New Hampshire leads at about 6 gal. (23 L), and on the other end of the scale is West Virginia at 1.3 gal. (5 L).[59] Total US retail sales of spirits in 2018 were valued at $92 billion.[60] Vodka accounted for about 30% of sales, with about 73 million (2.4 gal., 9 L) cases sold.[61]

What Is Changing for Spirits?

In the United States, bourbon—the homegrown spirit—must be made with not less than 51% corn. It must age for at least two years in charred new oak barrels in unheated barrel houses, where the oak imparts a characteristic dark amber color and the barrels cool and warm with the outside temperature, moving the bourbon in and out of the oak pores and extracting flavors.

What will climate change do to bourbon? While bourbon can be made in any state, Kentucky's hot summers, cold winters, and plentiful oak forests make it a near-perfect place. However, wetter conditions and more extreme weather will increase risks to grain production there as in other regions. Distillers can overcome this risk by depending not on a single source of corn that might be hit by climate change but on multiple sources to ensure a supply every year. Warmer conditions may also affect bourbon's flavor, and it is fairly certain that the "angel's share"—the amount of bourbon lost because of diffusion through the barrel—will increase as temperatures continue to rise. Climate change will also lead to a slight shift north in oak habitat, but this will be a slow process, and there should be plenty of oak available for barrels in the coming decades.[62]

How about Scotch and climate change? Scotch is whisky (without an "e") produced in Scotland, made mostly from malted barley and aged three or more years in oak barrels. Distillers who produce Scotch will face several challenges in the future, namely extreme weather, intense rainfalls, and less spring meltwater, which will reduce water availability and threaten grain supplies. The age-old malting, distilling, and aging processes are all temperature dependent and will be changing, affecting the entire Scotch-producing business.[63] In addition, characteristics of the stream water used in the malting process—which gives some Scotch brands a unique character—will likely change. Because of a drought in 2018, some distilleries had to halt production.[64] Last, as with bourbon, rising temperatures will permit the angel to take a larger share, more than the historic average of 2%, which translates currently to about 29 million gallons (110 million liters) per year.[65]

Think now about an icy cold margarita in a salt-rimmed glass. It is the most popular mixed drink in the United States, thanks in large part to Jimmy Buffet's song "Margaritaville." About nineteen million cases of its main ingredient, tequila, were sold in the United States in 2018.[66] Tequila is made from blue agave, also known as the century plant, which grows into a 6 ft. (1.8 m) tall, spiky, fleshy-leafed plant that is relatively hardy. The plants do well in the dry climate and volcanic soil of Jalisco, Mexico, where 95% of the world's agave is grown. After the plant is six to ten years old, harvesters remove its heart—which looks like a giant pineapple and can weigh up to 200 lb. (91 kg)—by hand, using tools that have been around for over one hundred years. The heart is then heated and crushed or pressed to release a sugary liquid, which is fermented and distilled. The tequila may be aged for a few months to a few years, or not at all.

In 2019, distillers in Mexico produced almost 93 million gallons (352 million liters) of tequila, with the United States importing most of it.[67] But rainfall variability, more snowfall and frosts, and temperature changes are threatening blue agave. In 2016, rapid snowfall and cold temperatures killed millions of plants, resulting in nearly a sevenfold increase in the price of agave.[68] When temperatures rise, plants grow too fast, decreasing the sugar needed for fermentation.

Now let's dissect another mixed drink and determine how climate change will affect its ingredients. A Bloody Mary is typically made with vodka, tomato, lemon, and lime juices, horseradish, garlic powder, Worcestershire sauce, Tabasco sauce, celery salt, and ground black pepper.

Let's start with vodka, the best-selling spirit in the world.[69] Unlike other spirits, vodka can be made anywhere from grains, beets, fruit, honey, or potatoes. The source of water can sometimes give a vodka a unique character. Although unflavored vodka is often used in Bloody Marys, natural flavorings can be added to vodka, many of which are threatened by climate change. As for the other ingredients, tomatoes are sensitive to high temperatures and yields may decline about 10% toward the end of the century in California.[70] Worcestershire sauce is made with anchovies, and with the warming of the oceans, populations of anchovies are likely to move to higher latitudes and deeper waters where conditions are cooler.

Such shifts have already occurred in the North Sea. Changes in the avail-ability of nutrients and ocean acidification may also affect anchovy popu-lations.[71] The cloves in Worcestershire sauce are also in trouble because cyclones have become more frequent and intense in Madagascar, putting clove trees at more risk.[72]

The McIlhenny family of Avery Island, Louisiana, has been making Ta-basco sauce for 150 years in the face of hurricanes and storm surges, such as Hurricane Rita, which came close to flooding their production facili-ties in 2005. They have invested in a 20 ft. (6 m) tall levee, a pump system, and backup generators to help ensure their annual production of 750,000 bottles of Tabasco sauce continues.[73] They are also restoring wetlands to help address the increasing threat of sea level rise. The tenacity and inge-nuity of the McIlhenny family should give us hope that the Bloody Mary will be around for a long time, despite the many changes.

Many distilled spirits are flavored and colored with botanicals, includ-ing nuts, fruits, roots, and a very long list of herbs and spices. An extreme example is Liquore Strega, a delightful Italian liqueur obtained by the dis-tillation of about seventy herbs and spices, including mint, fennel, and saf-fron. Many of these herbs and spices are commercially grown, but others grow wild in remote areas. Although it's nearly impossible to describe the impact climate change is having or will have on each unique ingredient as conditions warm, many of these plants might flower out of synchrony with insect pollinators and fail to produce fruit and seeds. If they grow in moun-tainous regions, their ability to move upslope is not guaranteed, and overall growing conditions are worsening with more weather extremes. Other plant species better adapted to the new conditions might overwhelm and displace these species, endangering the flavors and other important charac-teristics distillers have depended on for centuries.[74]

Consider saffron that comes from crocuses grown in India's Kashmir Valley. Changes in rainfall patterns and increasing incidence of drought have put this second-largest region of saffron production in rapid decline. Yields are dropping and so is the quality of this spice, potentially jeopar-dizing the livelihoods of thousands of farmers and many others involved in the saffron business.[75] Increasing extreme weather is taking a toll on many other spices, including turmeric, ginger, cardamom, and nutmeg.[76]

Overall, spirits as we know them won't be the same, and neither will economies and cultures around the world.

Prize that vanilla in your vodka

Each year the up-and-coming Corsair Distillery in Nashville, Tennessee, makes ten to twenty thousand cases of whiskey, gin, vodka, rum, and a few "oddballs," as head distiller Colton Weinstein calls them. Weinstein buys ingredients from all over the world to distill spirits such as their flagship Triple Smoke, a single malt whiskey made from barley and smoked with cherry wood, German beechwood, and Scottish peat. Fifteen botanicals flavor their Red Absinthe, including red hibiscus and tarragon. Citrus and vanilla impart essence to their Spiced Rum and Vanilla Bean Vodka.

He buys high-quality vanilla beans from Madagascar, which provides 80% of the world supply. "When you're selling vodka at a premium price," says Weinstein, "you have to use premium ingredients." Climate change—including a duo of cyclones—coupled with local bean theft and a boom-and-bust economy, drove vanilla prices in 2017 to $273 a pound ($600/kg). Recently, Weinstein says, he spent $3,000 or $4,000 for 10 lb. (4.5 kg), enough for about three months, depending on how much rum and vodka he makes with it.

What Is Being Done to Keep Our Spirits Up?

The distilled spirits industry is exceptionally diverse, as are its needs for many kinds of grains, vegetables, fruits, herbs, and spices. We discuss the challenges these foods are facing in more detail elsewhere in this book but offer some recommendations here.

To ensure a steady supply, the distilling industry will need to diversify its sources of ingredients. Switching to alternative varieties that tolerate the new conditions is another option. For example, some distillers are already making good tequila using commercially grown wild agave types, which seem to tolerate harsher conditions better than blue agave.

Many unique wild herbs and spices can't be readily grown under artificial conditions, so some will be lost as conditions change. We need more research to assess how to retain these unique sources of flavor.

The innovative spirits industry is leading the way to ensure that products remain on the shelf. Plant breeders are investigating a drought-tolerant agave cultivar that "breathes" at night instead of during the day, minimizing moisture loss. It may be possible to incorporate the key gene that controls this process into other agave plants, thereby increasing their ability to thrive under hotter and drier conditions.[77]

The distilling industry is also reducing its impact on the climate. Companies are switching to electric delivery trucks, increasing their use of solar and other renewable energy, and conserving water.[78] Bacardi has reduced its energy use by about one-third and its water use by over half since 2006, when the company started tracking its carbon footprint. Chase Distillery in Britain is aspiring to be carbon neutral by 2020, and the Diageo brands are reducing waste, carbon emissions, and water use.[79] Marble Distilling in Colorado runs a zero-waste facility that saves 4 million gallons (15 million liters) of water a year.[80]

Summing It Up

Just as the distilled beverage industry is diverse, so are the challenges a changing climate poses to it, notably risks to ingredients and changes in water quantity and quality. In the coming years these special beverages that many love and enjoy will face wide-scale change. Industry leaders will need to make deliberate, strategic decisions to keep these beverages flowing as risks intensify. The next time you raise your glass, consider also raising your voice in support of climate-change solutions.

Salads

Distinct, Diverse, Delicious

Whether you go for some mixed greens, a traditional Caesar, a classic iceberg wedge, or something of your own creation, the salad is a common way to start a meal. The crisp, tangy flavors stimulate the appetite. Of course, you can eat a salad at any point during the meal, or even *as* your meal, throwing in arugula, spinach, endive, kale, dandelion, radicchio, or lettuces and topping it with any vegetable, nut, cheese, fruit, meat, or seafood.

The word "salad" comes from the Latin word for salt: *sal*. The ancient Greeks and Romans started the trend of dining on raw vegetables seasoned with salt, vinegar, and oil, which they called *salata*, or salted things. *Salata* eventually transitioned to *herba salata*—salted herbs or plants. Hippocrates and Galen recommended eating salads at the beginning of the meal to aid digestion, whereas other ancient Greeks thought vinegar ruined the taste of wine and therefore believed that salads should be eaten last.[1]

Throughout the centuries, the salad evolved. The fourteenth-century French ate boiled watercress and chard with oil, cheese, salt, and meat broth, while their English contemporaries preferred raw leafy vegetables seasoned with parsley, sage, rosemary, chives, or garlic served with cooked meats. Although ancient physicians thought eating raw vegetables was healthy (provided they were counteracted with hot foods), green salads were not served until the late eighteenth century and only on upper-class dining tables.

In the United States, green salads were uncommon and even perceived to cause illness until the 1880s, when Emma Ewing published *Salads and Salad Making*. In 1893 Oscar Tschirky created the famed Waldorf salad of lettuce, apple, celery, and mayonnaise at New York's Waldorf Astoria

Family orchard adapts to the weather

She grew up riding pickup trucks around the orchard with her dad and helping her mom with data entry. During breaks from Dartmouth College, Jenny Crist Kohn would come home to plant. "I've always loved it, especially the field side of things," she says, recalling her dad placing trees in a furrow and workers adjusting the height, shoveling in dirt, and packing it with their feet.

A fifth-generation farmer, Kohn and her team manage Crist Brothers Orchards, a 550 ac. (223 ha) farm in Walden, New York, two hours north of New York City. They care for 381,000 trees—about twenty varieties in different shapes and sizes. Each year Kohn and eighty year-round workers make sure that half a million bushels of apples get properly picked, stored, packed, and shipped to outlets such as Walmart and Costco.

Kohn talks about her land, the trees, and what the weather will serve up. "We have seen strong weather events since I've been home on the farm," she says. June and July of 2018 were very dry, but the rest of the season it poured. She tackles both ends of that spectrum. For dry periods, they've installed pond-fed trickle irrigation networks that conserve water but also deliver what the trees need to grow apples to size. When it's wet, they depend on their well-drained slopes to prevent soggy soils. Water, however, is only one of their concerns.

"The danger of a spring frost is always really scary, and it seems like that's happening a little more," says Kohn. The orchard installed four 30 ft. (9 m) tall wind machines that draw warmer air down to the trees to prevent freezing.

Overall, Kohn thinks the current growing season is longer than in her grandfather's day. When her dad first experimented with Fuji apples, they worried the crop wouldn't mature before temperatures dropped below 25°F (-4°C) in the fall. Now Crist Brothers grows Fuji confidently and just expanded to Pink Lady, a variety that needs even later fall warmth.

(Continued on next page)

71

(*Continued from previous page*)

Of course, a longer season extends contact with diseases, insects, and more uncertain weather. "Hail seems to be relatively frequent in recent years and can destroy a crop really fast, in a minute," says Kohn. In 2018 the business invested in 10 ac. (4 ha) of netting that protects the trees against hail damage, bird pecking, and sunburn. "We're always looking for ways to protect our crop against disasters." Kohn would like to see more research on all areas of farm management. "Public funding for agriculture is really important because it's our food supply, and we have to stay ahead of it."

Hotel. On the other side of the country, where fresh ingredients were plentiful, salad creation became an art form. The well-known Green Goddess salad dressing was created at the Palace Hotel in San Francisco in the early 1920s. And shortly thereafter, in 1926, Robert Cobb introduced the Cobb salad at his Brown Derby restaurant in Los Angeles.[2] According to a 2018 survey, 83% of people in the United States either like or love salads.[3]

Like everything else on the menu, ingredients for both the salad and its dressing are changing. The balsamic vinegar used in many dressings is made from grapes, some of which can't thrive the way they used to, as the previous chapter detailed. Nuts and many fruits we use in salads are among the plants that need a winter cold spell or dormant period to develop, and as already mentioned the dormant period is shortening as winters wane. Apples in recent years have also been wiped out in parts of the United States by false springs.[4] The cheeses we enjoy in salads come from cows that produce less milk when they're hot.[5] The flavors of our favorite greens and vegetables may also change.[6] On the bright side, in some areas of Africa the production of sesame could expand because the crop tolerates the droughts that are becoming more common and adversely impacting traditional crops. Sesame could become an important source of income for small-scale farmers.[7]

Two salad fundamentals—avocados and olive oil—illustrate the challenges many salad ingredients face. Exploring how to keep these two available also lets us see how we can ensure our salad fixings remain plentiful.

Avocados

Enjoying slices of avocado in a salad is common today, and our love for this fruit has increased rapidly. In 2018, people in the United States ate over 8 lb. (3.6 kg) of avocado each, about four times more than in 2000.[8] The total annual consumption is close to 2.6 billion pounds (1.2 billion kilograms), and on Super Bowl Sunday, fans down 4,000 tn. (3,600 t), most mashed into guacamole.[9]

Consumers choose avocados because they believe they're healthy, contain "good" fats, and are tasty.[10] And this is supported by the evidence. One avocado has twice the potassium of a banana and offers vitamins C, K, B5, B6, and folate.[11] Avocados also add fiber and healthy fat to a salad, make great baby food, and can be blended into a creamy dressing or dessert.

Where Do Avocados Come From?

The avocado is indigenous to three regions with genetically distinct populations: the Guatemalan highlands, the Pacific coast lowlands of Central

America, and south central Mexico.[12] Today more than sixty countries grow avocados—or *aguacates* in Spanish—commercially.[13] Mexico is the world leader, producing over 2 million tons (1.8 million metric tons) in 2017, followed by the Dominican Republic and Peru.[14]

Hundreds of varieties of avocado exist—with names such as Bacon, Linda, Maluma, and Serpa—but the most common sold in the United States is Hass, named after a California postal carrier who became an avocado grower, Rudolf Hass. Hass avocados can be harvested almost all year long in some regions and make up 95% of all avocados sold in the United States.[15]

Although California, Florida, and Hawaii are responsible for almost all US avocado production—with California contributing the lion's share—they can't keep up with demand.[16] The United States is now the largest avocado importer in the world, buying close to $2.6 billion worth in 2017, mostly from Mexico.[17]

How Are Avocados Grown?

Most avocado trees are actually two trees in one. The top (scion), a variety with a distinct flavor, texture, and other marketable traits, is grafted onto a rootstock—the lower trunk and roots of an entirely different type of avocado tree. This rootstock is often selected for resistance to diseases.[18] The resulting seedling is eventually taken from the greenhouse and planted in the orchard (or as some say, the grove).

The avocado tree is picky about its habitat: the weather can't be too cold, windy, dry, or soggy. Cold temperatures can reduce the period of fertility.[19] For most varieties, even a light frost can cause fruit to drop early. During the flowering period, blossoms can handle light breezes, but low relative humidity can dehydrate blooms and high winds can knock them off the tree.

Avocados must have well-aerated soil and the correct amount of water to create their oil-rich delights. Every pound (0.5 kg) of mature fruit needs an average of 142 gal. (538 L) of water over the fourteen-to-eighteen-month growing period. If everything goes right, a mature tree in California can produce an average of 150 avocados (60 lb., 27 kg) a year.

Like bananas, avocados mature on the tree but ripen only after they are harvested. They are very hard when picked and can be transported long distances. Before being sold, many commercial avocados are exposed to ethylene gas to start the ripening process or stored at room temperature for a week or two before sale. Consumers can ripen avocados by placing them in a paper bag, which traps the natural ethylene.

How Large Is the Avocado Industry?

The industry is lucrative and growing rapidly, both in the United States and elsewhere.[20] Worldwide harvests of 6 million tons (5.6 million metric tons) placed avocados above kiwi, apricot, and cherry production by weight in 2018.[21] Retail sales of avocados in the United States in 2017 were $2.3 billion, up from $1.1 billion in 2014. More avocados means jobs in production, transportation, and sales. In Michoachán, Mexico, avocado exports grew 240% from 2011 to 2016. About 300,000 jobs there—a whopping 57% of employment—are linked to avocado production. Avocados reportedly reap more cash than any other Mexican crop, including marijuana, a fact not lost on the cartels benefiting from the boon and the locals who are unable to afford their own *aguacates*.[22]

What Is Changing for Avocados?

A warming climate is threatening avocado trees. United State growers must cope not only with droughts in the West but also with water-logging storms and hurricanes in the Southeast. California avocado yields could decrease about 40% by 2060 (compared with average yields from 2000 to 2003) owing to climate change, unless growers are able to adapt their production practices.[23] Future production might not be viable for several reasons.

First, avocados need lots of water. In the United States they're grown in regions such as San Diego County, where rainfall is often scarce, water distribution policies are complex, and irrigation depends partly on snowmelt or groundwater. Even with relatively efficient drip irrigation, an acre of mature California avocado trees requires about 4,500 gal. (17,000 L) of water a day, depending on the location, temperature, and wind.[24] Warming

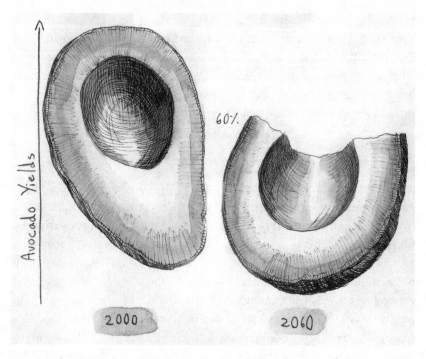

Because of climate change, California avocado yields in 2060 might be only 60% of what they were in 2000–2003. (David B. Lobell et al., "Impacts of Future Climate Change on California Perennial Crop Yields: Model Projections with Climate and Crop Uncertainties," *Agricultural and Forest Meteorology* 141, no. 2 [December 20, 2006]:214, https://doi.org/10.1016/j.agrformet.2006.10.006.)

temperatures exacerbate the issue of water availability by increasing water evaporation from the soil and plant leaves.

In addition to avocado farmers, hundreds of thousands of people in Chile and Peru depend on water from melting glaciers.[25] Scientists and many others worry that as these glaciers melt away in coming decades this source of irrigation water will decline, as will the production of avocados and many other crops.[26] Water activists in the Valparaiso region of Chile claim that 2.5 ac. (1 ha) of avocados demand the equivalent of what one thousand people need daily to survive—and there's not enough water.[27]

In California, the use of recycled or desalinized water seem like possible solutions. At least one study, however, shows that switching from conventional irrigation to recycled water would increase avocados' life-cycle greenhouse gas emissions by 9%. Using desalinated water would increase emissions by 270%—a dubious prospect.[28]

Temperature fluctuations are another challenge. During fruit set (usually winter through early spring), Hass avocado flowers need nighttime temperatures between 50° and 54°F (10°–12°C) to develop properly and ensure good yields.[29] Daytime temperatures during the fall, which determine fruit set, are also critical.

Establishing new orchards in cooler latitudes is probably not realistic. In countries that currently grow avocados, much of the suitable land at higher latitudes is already occupied, and even as new trees might become established and start producing, climate changes would be "catching up" in those areas.[30]

Increased global demand has spurred the removal of indigenous forests to make room for avocado production. Global Forest Watch estimates losses in some areas of Mexico to be in the range of 15,000 to 20,000 ac. (6,000 to 8,000 ha) per year.[31] Young avocado orchards can't mimic complex indigenous arboreal systems, and even older orchards, with their upper and lower stories of growth, are not as ecologically diverse as the forests they have replaced. In addition, mature avocado orchards require nearly twice the water similarly sized forests do.[32]

Extreme weather could continue to take a bite out of US avocado production. In 2017 Hurricane Irma, with winds exceeding 100 mph (160 kph), dumped up to 11 in. (28 cm) of rain in parts of southeast Florida, destroying about 40% of the crop.[33]

Worldwide, dozens of insects and at least forty diseases and disorders afflict avocados.[34] One extremely damaging disease is *Phytophthora* root rot, common in water-logged soil.[35] To minimize this risk, Hass growers use plants with disease-resistant rootstocks, establish good drainage and adequate organic matter in their groves, and irrigate carefully.[36]

In Mediterranean-like climates, the persea mite weaves silken nests on the undersides of avocado leaves, where it feeds and reproduces, damaging the leaves. Two species of helpful predatory mites feed on the persea mite, but they don't like the kind of hot, dry summers that California experienced in 2008 and 2009. Studies indicate that future weather patterns might doom any effective control from these predators.[37]

South American avocado growers face several pests, such as the carambola fruit fly, which attacks not only avocado but also cashew, papaya, guava, and citrus fruits. About 20% of Brazil's avocado acreage is grown in areas where this fruit fly could thrive as the climate changes.[38]

What Is Being Done to Save Our Guac?

Several international programs are tackling deforestation in general, including deforestation resulting from production of avocado and other crops. Among them are the United Nations' Food and Agricultural Organization's Reducing Emissions from Deforestation and Forest Degradation program and the Rainforest Alliance.[39] By 2017 the Rainforest Alliance had certified 8.6 million acres (3.5 million hectares) of forest worldwide, including avocado farms.[40]

Where feasible, avocado growers can switch from irrigating for a set number of hours to a "Smart Irrigation" system. Smart Irrigation considers soil moisture content, local weather, and estimates of water evaporating from the soil and released from the tree. Researchers in Florida have simulated how switching to the Smartirrigation phone app reduces water inputs about two-thirds.[41]

Using dwarf or semidwarf avocado rootstocks can produce smaller trees that could perform better than those with conventional rootstocks. Some of these new trees might benefit from increased carbon dioxide in the atmosphere, and others might absorb nutrients more readily.[42] A California technique called high-density planting can double the avocado yield with less water than conventional methods but only for certain cultivars and only if trees are kept small.[43]

Replanting an orchard with new varieties takes time and money. One compromise might be to "top-work" existing trees, that is, regraft their rootstocks with varieties that tolerate warm temperatures. Finally, as a word of encouragement, some of the nine hundred named avocado varieties are already adapted to subtropical and tropical conditions and might handle climate change better than Hass and other current favorites.[44]

Olive Oil

From hard green drupes
of bitter flesh, a river
of gold and green

BARBARA CROOKER, "ODE TO OLIVE OIL"

We dress our salads with olive oil, massage it into our hair and skin, make soap, lip balm, and candles from the liquid gold, and even use it to polish furniture. Olive oil is versatile, popular, and, in some areas of the world, imperiled.

What Are the Origins of Olive Oil Use?

More than six thousand years ago, inhabitants of the Mediterranean Basin began harvesting olives from wild trees. Residents of what is now Syria figured out how to press oil, and the idea spread to the neighboring lands—today's Israel, Spain, Greece, and Italy. Records show that for millennia, olive oil was used for lamp oil, soap, ointments, and religious anointing but not for cooking.[45]

An olive tree can live hundreds of years, and thousands of varieties exist worldwide. Some varieties produce table olives that are cured and eaten; others produce olives for olive oil. Many Mediterranean growers have small plantings of region-specific varieties that have contributed to the nineteen classic styles of olive oil. Slowly these growers update varieties when they plant new groves, but growers in countries where new groves are burgeoning are the primary sources for the two dozen cultivars now popular worldwide.[46]

Where Does Olive Oil Come From?

Olive trees thrive where summers are warm and dry and winters are cool and wet. Spain is the largest producer of olives, followed by Italy and Greece; the combined production from these three countries totaled 14 million tons (13 million metric tons) in 2018.[47]

California is responsible for most of US-made olive oil but contributes less than 1% to global olive oil production. Because people in the United States love olive oil, they import more, by volume, than any other country in the world.[48]

On average, European Union countries produce most of the world's olive oil. ("Olive Oil Production in Selected Countries Worldwide from 2011/2012 to 2017/2018," Statista, released April 2019, https://www.statista.com/statistics/192606/world-olive-oil-production-in-selected-countries-worldwide/.)

How Are Olives Grown?

Olive trees tolerate periods of hot, dry conditions, so growers traditionally plant them on nonirrigated hillsides where other crops won't thrive
(especially in the Mediterranean) and typically wait three years for the first
fruit. Like apples, citrus, and several nuts, the olive tree is alternate bear-
ing: a highly productive year is followed by one much less productive.

Young green fruits turn purple as they mature, black when fully ripe.
The olive variety and the stage of maturity when the fruit is pressed determines how olive oil will taste.

How Is Olive Oil Produced?

Traditionally, workers pick the fruit by hand or beat the branches with
sticks. Mechanical harvesters that either shake limbs or agitate the tree
canopy are also widely used. These canopy-contact harvesters require trees
planted and pruned specifically for this harvest method, with 200 to 900
trees per acre (500 to 2,200 per hectare), in contrast to the 70 or fewer trees
per acre (173 per hectare) for traditional hand-harvested groves. Modern
groves planted with short, compact trees on gently sloping ground can also
be harvested with machinery similar to mechanical grape harvesters.[49]

Table olives are too bitter to eat straight from the tree, so they need to
be cured with water, brine, or a lye solution. California-style green olives
have been soaked in lye until it penetrates all the way to the pit; an iron
compound is added to fix a black color. Spanish style olives are soaked in
a lye solution that penetrates partway; then they are fermented in brine.
Olives can also be cured in brine alone, without lye, which takes longer.[50]

Olives destined for oil are usually washed and then crushed in a mill to
form a paste, which is kneaded at a warm temperature to release the oil.
Traditionally the paste would be pressed to extract the oil, but today most
processors centrifuge it, put it through a separator to remove any remaining water or solids, and filter it.

Virgin and extra virgin olive oils are made mechanically, without sol-
vents or high heat.[51] Their pungent, bitter flavors, which indicate a richness in antioxidants, add complexity to many dishes. These oils command
premium prices compared with lower-quality olive oil marketed in the

United States as Light, Extra Light, or even Pure. Such refined oils have typically been treated with charcoal, chemicals, or heat.[52]

Who Cares about Olive Oil?

You do, probably. From 2000 to 2018, US consumers boosted olive oil use by 60%, enjoying 370,000 tn. (336,000 t) in 2018.[53] Globally, the market value of olive oil was $11 billion in 2018 and is expected to increase to $17 billion in 2028.[54]

Why such popularity? When asked in a recent survey, almost 90% of US consumers claimed that olive oil is healthy, and research is beginning to back this up.[55] By consuming it regularly, you could live longer because it helps to protect the heart by decreasing risks from diabetes, metabolic syndrome, and obesity. It has also been implicated in preventing some cancers.[56]

Growers certainly care about the olive-oil industry. One reason is that the microeconomics of olive oil make sense when the weather is cooperating. Mediterranean groves are often small and the ownership diverse, generally allowing for financial independence. Furthermore, olives are harvested in the winter months, providing economic opportunity when other agricultural activities in the region are slow paced.

What Is Changing for Olives?

Drought. While olives can tolerate periods without water, extended droughts can decrease vegetative growth, yield, olive size, and the oil content of the fruit. Experimental trees damaged by drought lost half of their ability to convert sunlight to sugars; even after they were finally watered, they never fully recovered.[57] Droughts during 2012 and 2014 stressed the trees in Andalusia—the region responsible for 80% of Spain's olive oil and nearly one-third of the world's production.[58]

Summers in the Mediterranean will be hotter and drier than in the past, with warm temperatures extending into spring and fall. The need for irrigation is expected to increase 18% over the region.[59] Groves that are irrigated can produce up to ten times more fruit than can groves dependent on natural rainfall.[60] Irrigation is an option for some groves, but systems are expensive and groundwater must be available.

Erratic weather. Coupled with drought, erratic weather is challenging olive production, especially for the nonirrigated, desert-like groves most vulnerable to crop loss.[61] In Italy, early spring frosts, temperatures above preindustrial levels, and a general decline in precipitation resulted in a 57% drop in 2019's olive harvest. Climate-related events across the Mediterranean that year resulted in the poorest regional harvest in twenty-five years.[62] In Spain, shifts in weather patterns caused by climate change are resulting in more freeze injury to the fruit and subsequently a "frostbitten flavor," now one of the most common defects in olive oil quality.[63]

When climate fluctuations weaken olive trees, they become susceptible to insects and diseases.[64] One of these is a deadly bacterium that infects fruit and nut trees, which has already killed a million olive trees in Italy. It arrived in Spain in 2017.[65]

Higher temperatures year-round. Warmer winters are threatening the development of the buds that bloom and form olives because, much like avocados, olives are facing fewer winter chill hours. California is already seeing a decrease in winter chill hours, and scientists predict growers will experience an average of forty fewer chilling hours per decade.[66] Most groves in southern Europe are also expected to have fewer chilling hours in the future.[67]

Too much heat in late winter is equally problematic if it causes an early bloom, when an abrupt frost can thwart pollination. That's exactly what happened to California olive growers in 2018: a false spring that led to a 50% drop in olive oil production for the season.[68]

The warming climate also affects the olive fruit fly.[69] During rainy, balmy weather, this pest breeds prolifically and deposits its eggs inside the olives, where they hatch into small worms (larvae) that feed on the fruit for about a month.[70] Conversely, dry, hot weather can kill olive fruit fly eggs, worms, and even adults if they can't find food and water. That turns out to be a consolation prize for growers dealing with drought.[71] As the climate warms and groves are being planted in more northerly regions, however, this fruit fly is tagging along; studies show it is migrating northward from its southern range.[72]

What Is Being Done to Keep Olives Growing?

Growers of other crops have developed an irrigation method that could work in super-high-density olive groves. Regulated Deficit Irrigation relies on the idea that olive trees can tough it out for most of the year except during certain growth stages, such as flowering. If rainfall is inadequate, growers would irrigate only during those times, reducing their water use.[73]

Breeding new olive cultivars to match an unpredictable climate is in its infancy but could help growers immensely. In arid regions, trees that flower early—when rains are more likely—would theoretically set fruit better than late-flowering cultivars. Developing new cultivars with lower winter chill requirements could insure a good crop even as winters warm.

Growers and scientists are also working on other options to mitigate the effect of climate change on olive groves. For example, updated modeling would help growers predict when olive trees will flower and thus when irrigation and other interventions would be most effective. The new AdaptaOlive simulation model for semi-arid conditions takes into account climate change and water requirements, then forecasts olive tree responses, yield, and economic impacts.[74]

Some Greek olive growers are successfully growing cover crops such as common vetch, barley, or chickpeas between rows of olive oil trees. The leguminous crops return nitrogen to the soil, reduce the need for added fertilizers, provide food for animals and humans, and bring in extra income.[75]

Climate change and increasing demand are already shifting olive trees' home turf. In the Mediterranean, groves are predicted to expand 25% by 2050, but farther north and at higher elevations.[76] In the United States, Oregon farmers began growing fifteen varieties on about 50 ac. (20 ha) in the Willamette Valley in 2017.[77] It's a small but significant start because these groves, with adequate water and moderate temperatures, are much farther from the equator than their Mediterranean counterparts. Areas of the United States, Argentina, and some other countries could continue to gain importance as olive-growing regions because of adequate water and moderate temperature.[78]

Extend an olive branch to new cultivars

"One of the most noticeable impacts of climate change," says Dan Flynn at the University of California, "is that the harvest is starting earlier than it used to." Flynn is executive director of the University of California–Davis Olive Center, which focuses on education and research for California olive growers and processors. When the center opened in 2008, growers began harvesting olives in late October and early November. Now they might start in early October—an indication of warmer seasons speeding up the calendar for bloom and development.

Citing heat waves in the fall and frigid weather when the tree is ready to bloom, Flynn says, "The transitions between seasons are more erratic than they used to be, with abrupt rather than gradual changes." A freeze before harvest can make the fruit decompose quickly, ruining the option of extra virgin olive oil, and a sudden freeze before dormancy sets in can defoliate or even kill trees.

Flynn thinks growers should supplement calendar-based farming with data-based farming. They can use instruments to track temperature and moisture, assess how weather patterns are changing, and adjust their practices.

Growers can also plant trees better adapted to the heat. "There are some varieties of olives more suited to the desert climate, and we should do trials on those in California," says Flynn. Past California research concentrated on Manzanilla—the primary cultivar of the California-style ripe olive and also the ubiquitous green Spanish-style table olive with a pimiento stuffed in it. Flynn thinks research should focus on cultivars that don't have a long chill requirement. Of the hundreds of varieties of olives worldwide, large-scale producers use only a few dozen. "There are millions of acres of varieties in Turkey that we don't have experience with in California," he says. "Maybe we better be looking at them if they're adapted to a warmer climate."

It's not easy or quick to find varieties that will adjust to California's climate, resist diseases, and lend themselves to efficient harvesting. With traditional breeding methods, developing a new cultivar will take a long time. Says Flynn, "We should at least be exploring it and making sure we have some options down the road."

It's not all bad news

Could climate change ever *benefit* olives? In some cases, yes. Scientists in Spain predict that higher levels of carbon dioxide in the future will generally negate the effects of climate change on olive production, except where there is a lack of adequate water. If in areas of Spain growers adapt how they manage water, they could actually increase yields by the end of the century.[a] Warming of colder areas (those that were previously too cold to grow olives) will likely lead to expanded production to the north in the Northern Hemisphere and to the south in the Southern Hemisphere.[b]

---

[a] I. J. Lorite et al., "Evaluation of Olive Response and Adaptation Strategies to Climate Change under Semi-arid Conditions," abstract, *Agricultural Water Management* 204 (May 31, 2018):260, https://doi.org/10.1016/j.agwat.2018.04.008.
[b] Paul Vossen, Vossen Ag Consulting, Santa Rosa, CA, personal communication, March 9, 2019.

Summing It Up

Salads today are more diverse than ever before, thanks to the bounty of fruit, nuts, greens, and herbs grown around the world plus amazingly complex dressings. As the climate changes, so do the opportunities to keep this menu item well stocked. Scientists are developing more resilient varieties, farmers are adopting new methods to manage scarce water supplies, and some production will move to more hospitable regions. Over time, consumers may need to adapt to the loss of some salad ingredients and to unique flavors and textures as new salad ingredients come on line. We will need to work together to keep this delicious item on the menu.

Our palate now pleased with drink and salad, we move to the main course.

The Main Course

When we gather around the table, one dish, the main course, is often central to the meal. We plan our side dishes around it, we pair our beverages with it, and our eyes light up when it finally arrives on the table. Typically, that main attraction includes animal protein, and as incomes continue to rise around the world, so, too, will the demand for meat. Climate change is affecting the animals we eat—and some of the animals we eat are also driving climate change.

Meat's long evolutionary history with humans goes back some 2.6 million years. The addition of meat to our ancestors' diet likely helped us develop our large and complex brains and become highly social primates.[1] Today, people in the United States eat a lot of beef, lamb, chicken, pork, and fish because for many they are delicious and can be prepared in so many different ways. Steak, lamb shank, roasts, chorizo, goulash, and hamburger may constitute the main dish, or we may add meat or fish in small amounts to pasta dishes, rice, or other grains. Meat provides high-value protein, with all the essential amino acids children and adults require, and it contains critical fatty acids and several micronutrients, such as iron, zinc, and vitamins.[2] The United Nations' Food and Agriculture Organization (FAO) considers the livestock sector important to global sustainable development goals because it contributes to national economies, reduces poverty, helps end hunger and malnutrition, and provides nutritional benefits that enhance children's cognitive development.[3] However, in developed countries, consumption of meat—particularly red meat—often exceeds dietary recommendations.

Beef: Big Business, Big Scale

The United States is the world's largest producer and consumer of beef. In 2018, US sales of beef were approximately $67 billion.[4] Annually, people in the United States consume about 57 lb. (26 kg) of beef per person.[5] In 2018, people ate 67 million tons (61 million metric tons) of beef worldwide, with the European Union, China, and Brazil top consumers at about 8.8 million tons (8 million metric tons) each year.[6] Beef production is a big business that requires large expanses of grazing lands. In the United States, this amounts to about 790 million acres (320 million hectares), or over 40% of the country's continental land area.[7]

Many of us consume beef, and savor it. It is essential to the lives of millions around the globe, but it also has a large carbon hoofprint. Beef, dairy cattle, sheep, and goats are ruminants, which means they are able to consume and digest coarse grasses, stalks, and husks—something that humans, pigs, and chickens cannot do. They accomplish this in part by enteric (intestinal) fermentation of these plants in the microbe-rich first compartment of their stomachs. This fermentation produces methane that the cow releases via burps. Beef produces about fifty times more greenhouse gas emissions per standard unit of protein than wheat and six times more than pork.[8]

Beef and mutton also require a great deal of land area to produce a standard unit of protein compared with most other food types. Grains

Not just milk and beef

When we think of cattle, we think of meat, but according to the US Department of Agriculture approximately 44% of the animal is used for byproducts such as luggage, medical supplies, cosmetics, soaps, and gelatin, to name a few.[a] Gelatin, widely used, is made by boiling the skin, bones, and cartilage of cattle and is found in gummy candies, marshmallows, sauces, desserts, and soups. It is also consumed for such health benefits as managing blood sugar, easing joint pain, and losing weight.[b]

---

[a] D. Marti, R. Johnson, and K. Matthews, "Beef and Pork Byproducts," USDA Economic Research Service, September 1, 2011, https://www.ers.usda.gov/amber-waves/2011/september/beef-and-pork-byproducts/.

[b] MaryAnn De Pietro, "10 Health Benefits of Gelatin," Medical News Today, accessed February 13, 2020, https://www.medicalnewstoday.com/articles/319124.php.

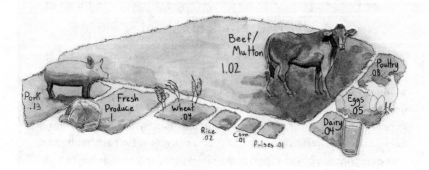

Contrasting amounts of land needed to produce the same amount of protein from plant- and animal-based foods. Pulses are dry edible beans, chickpeas, peas, and lentils.

such as corn, rice, and wheat use only 1% to 4% as much land as beef or mutton. Poultry and pork use about 8% and 13%, respectively, as much, suggesting that even shifts to other types of meat is helpful from a climate and land-use perspective.[9]

There are several ways to raise a beef animal, but typically a calf stays with its mother and nurses and grazes for six to ten months. Weaned calves may continue grazing or move to a feedlot where their diet shifts to high-energy grains, grain by-products, vitamins, minerals, and roughage (such as hay) for another two to six months. For their final four to six months, all commercial cattle consume a high-energy grain-intensive diet in feedlots. Animals raised this way are considered feedlot or grain-finished beef. Their diet increases their growth rate, helps ensure a heavier animal at harvest, and provides the quality of meat consumers demand, efficiently and at the least cost.

According to the National Cattlemen's Beef Association, there are over thirty thousand feedlots in the United States.[10] Of these, about 5% have one thousand head or more, and these larger feedlots account for about 80% of the fed animals.[11]

Beef production occurs in every US state, but over 70% of the feedlots are concentrated in Nebraska, Texas, Kansas, Iowa, and Colorado, where rangelands are replete with native grasses and vegetation for grazing cows and calves. In addition, producers there can readily buy grains and grain by-products for finishing cattle.[12] For example, Iowa alone produced 2.5 billion bushels of corn in 2018.[13]

Farmers also raise beef entirely on pasture and other forages. Such beef, typically labeled grass-fed, accounts for a small fraction of US-produced beef. In reality about 75% of grass-fed beef consumed in the United States is imported. Retail sales of grass-fed beef, however, have been doubling each year recently, and consumer interest in the health and environmental benefits of this type of production is expected to increase demand further.[14]

Views differ on whether grass-fed or feedlot-fed cattle have a greater impact on climate change. It's complicated. Because grass-fed cattle generally graze on forages of lower nutritional quality, they take longer to reach the desired weight, so they live longer. Consequently, they release more methane over their lifetimes than cattle finished in feedlots with high-energy grains. Scientists estimate that making a complete shift to grass-fed beef while meeting current demand would require a 30% increase in the number of cattle. They also suggest that transitioning away from feedlot production to an entirely grass-fed system could result in about 43% more methane emissions per animal.[15]

Others suggest that cattle could graze on the millions of acres of land that are generally unfit for cultivated crops. And somewhat like the buffalo that roamed the plains and returned their waste to the soil, cattle can

Comparison of how grass-fed (top) and feedlot-finished beef cattle (bottom) are raised.

When the grass isn't greener on the other side

"Don't panic, adapt," says rancher Pat O'Toole. And he lives by this motto, too.

A former representative in Wyoming's state government, O'Toole serves on several boards that promote agricultural adaptation, including Partners for Conservation and Solutions from the Land. He and his wife, Sharon, own Ladder Ranch in southern Wyoming, where the Rocky Mountains fringe grassy plains and the Little Snake wends its way to the Colorado River. "It's an incredibly beautiful place," says O'Toole of the homestead that has been in Sharon's family since 1881. "The soils are all volcanic and productive. The grass is really good, and the animals do really well."

He means sixty-five hundred sheep, a thousand head of cattle, forty livestock guardian dogs, thirty Border Collies, and sixty horses—many adopted through the Wild Horse Program at Wyoming's Department of Corrections.

Ladder Ranch manages grazing on its own land and on government allotments. In normal years, rain and snow make the national forest lush and more resilient than the unforested areas. In the spring, Peruvian sheep-herders bring the Ladder Ranch sheep to these high-altitude forests, then trek them 150 mi. (241 km) in the fall to lower-altitude winter pastures in Wyoming's Red Desert.

Frequently the O'Tooles rotate ewes and cows through the same landscape. The sheep browse primarily on sagebrush and broadleaf plants; the cattle prefer the grass. "If you do both in the same landscape correctly," says O'Toole, "you get 40% more utilization. That means roughly 40% more pounds of meat."

Ladder Ranch raises sheep for both wool and meat, and O'Toole tries to graze them year-round on fresh grass. Every December and January he supplements with corn, which keeps them strong and adds pounds, helping the ewes to deliver twins. When winter snowmelt is scarce or the previous season so dry that grass doesn't grow, he must feed corn longer. "Twenty eighteen was one of the driest that anybody's recorded," he says. "Some places were 70%–80% below normal." In 2019 O'Toole fed his sheep until April—at 3,000 lb. (1,360 kg) of corn a day, plus alfalfa. "It's more expensive," he says, citing the harsh winter as another challenge the sheep had to surmount.

To keep the land as resilient as possible, Ladder Ranch didn't graze parched areas. "When you have just one place, you don't have much flexibility. If you have several landscapes, you can prepare for the worst-case scenario and have more capability to survive."

Since 2005 O'Toole has been president of the Family Farm Alliance, an organization representing irrigators in the seventeen western states. This year farmers and ranchers in his region didn't have enough water and had to assert their water rights. "We're drilling wells and putting in solar pumps to duplicate springs that we used to count on." Ranchers, communities, and the state are also building reservoirs to catch water earlier. They know storage is going to be important and are adapting to trends. "To me," he says, "the future is about resiliency."

In 2014 the O'Tooles received Wyoming's Leopold Conservation Award. They currently protect habitat for grouse and other birds and plan to regenerate forests devastated by the bark beetles that have thrived on decades of warmer weather. "We have hundreds of miles of dead forest. You can't ride into the trees because they're all falling down, and the watershed itself is not functioning." O'Toole's family hopes that controlled burning, thinning noxious weeds, and regenerating aspen will make a difference. "I may be too old, but maybe my grandkids can ride through the trees again like we used to."

be part of a healthy system if the land and cattle are managed correctly. This approach is often referred to as regenerative agriculture or managed grazing. One group of scientists studied a grass-fed system in which the animals grazed on well-maintained high-quality forage and were finished on high-quality purchased alfalfa. Under these conditions, the plants and soil sequestered enough carbon to offset the emissions from the cattle. These scientists suggest that carefully managed grazing could help mitigate climate change.[16] Project Drawdown, an organization that identifies the most viable climate change solutions, also affirms that managed grazing systems can offset emissions from the animals at least until the soil becomes carbon saturated.[17] In contrast, the Food Climate Research Network concludes that at a global level, the level of carbon sequestration is small and outweighed by the greenhouse gases that the animals emit.[18] It's a topic requiring ongoing research and development across diverse landscapes and one with great potential impact given that grazing occurs on about one-quarter of the world's land area.[19]

Another strategy for raising beef and dairy animals is silvopasture. "Silvo" is Latin for forest, and silvopasture consists of trees and generally a forage crop. Silvopastures that include 20% to 30% trees capture far more carbon than open pastures. Manure from the grazing animals plus nitrogen-fixing forages planted among the trees improve soil and tree nutrition; the grazing also reduces brushy species, which are a potential fire hazard in some regions.[20] This approach is so good at capturing carbon that it is one of the top ten Project Drawdown climate solutions. Drawdown experts estimate that silvopasture could expand from the current 350 million acres (142 million hectares) to 550 million acres (223 million hectares) worldwide, reducing carbon dioxide emissions by 31 billion tons (28 billion metric tons) by 2050. In addition, silvopasture protects cattle from severe weather and high temperatures, and it lets farmers make additional income if they grow nut or fruit trees. Silvopasture requires time to establish but, where conditions permit, presents many opportunities.[21]

What Is Changing for Beef?

Climate change is affecting beef production, and increasing temperatures have the greatest impact. In addition, changes in rainfall amounts and patterns affect the quantity and quality of forage feed for all livestock. When

Tough times, tough decisions

A changing climate continues to barrel down on our food system and at the same time the world requires more food for a growing population. The changes needed are at a global scale, and they encompass economic, social, environmental, moral, and political issues. For example, people in rich countries should consume fewer animal products, but cutting back would affect tens of thousands of people who now raise, transport, package, and sell those products. Elsewhere around the world animal products are vitally important to the lives of millions. And growing crops in vertical farms may produce more of the food we need but comes with a higher carbon footprint. Experts assert that we need to revamp nearly every part of the global food system to battle climate change and nourish the world.[a] We need to make difficult trade-offs. This book addresses some of these issues, but many remain.

---

[a] Cynthia Rosenzweig et al., "Climate Change Responses Benefit from a Global Food System Approach," *Nature Food* 1, no. 2 (February 2020):94–97, https://doi.org/10.1038/s43016-020-0031-z.

Too much red meat—or too little

In January 2019, Lancet and the nongovernmental organization EAT published a report on the optimal diet for all people on the planet. The analysis takes into account not just what we can eat for health but also what fits within the limits of the planet's resources. The diet balances the food groups, rather than eliminating any of them. According to the report, people in the United States eat much more red meat than is considered healthy, while those in other regions around the world are not eating enough red meat.[a] For the health of people and the planet, a shift to a more plant-based diet in rich countries is highly desirable—consider beef a delicacy rather than a staple.

---

[a] Walter Willett et al., "Food in the Anthropocene: The EAT–Lancet Commission on Healthy Diets from Sustainable Food Systems," *Lancet* 393, no. 10170 (February 2, 2019), https://doi.org/10.1016/S0140-6736(18)31788-4.

Where's the beef?

Consumers have eaten plant-based meat alternatives—such as veggie burgers made of black beans, nuts, chickpeas, grains, tofu, mushrooms, and many other ingredients—for some time. Now, however, people looking to reduce or stop their meat consumption have additional no-meat versions of the foods they love.[a] Products such as Quorn are made from a single-celled protein fermented from fungi. Beyond Meat's veggie component is pea protein.[b] The Impossible Burger uses a soy-based heme, an iron-rich molecule, to mimic the taste of a traditional meat hamburger with a much smaller carbon footprint.[c] Plant-based burgers, sausages, and chicken-like nuggets are widely available in stores across the United States and in Europe, as well as in some fast-food chains. In the United States, the retail value of plant-based meat substitutes exceeded $800 million in April 2019 and the size of the market increased 37% between 2017 and 2019.[d] The company Beyond Meat drew significant investments when it offered publicly available stock in May 2019.[e]

Cultured meat and fish are also possible components of a low-carbon diet. Culturing adopts the method used in the medical world (to grow human tissue) to growing edible animal protein. The process involves using stem cells from soy, beef, chicken, or fish in a bioreactor that contains the necessary nutrients to promote cell growth and reproduction. These cells develop in an enriched growth medium that generally contains amino acids, antibiotics, growth serums, and glucose for energy.[f] The result is a meat product generated without the need for a living animal. Studies suggest that the production of cultured meat would produce up to 96% less

temperatures rise, cattle consume more water, lose their appetites, become less fertile, and can even die.[22] Scientists estimate heat stress costs the US beef industry about $370 million per year.[23]

The US Department of Agriculture (USDA) assessed how many US cattle were in regions experiencing drought in 2008, 2013, and 2015 and found it was 26%, 73%, and 12%, respectively.[24] Droughts that reduce the availability and quality of pasture, harvested forages, and grain for cattle are expected to increase and intensify in future years. Texas has experienced a series of droughts forcing farmers to reduce their cattle herds.[25] Some cattle producers have to transition their animals from grazing to feedlots sooner than normal when grazing lands deteriorate during droughts.[26]

greenhouse gas emissions and would require 96% less water and 99% less land than conventional production of meat.[9]

Although the technology exists to produce beef and other animal products this way, being able to buy cultured meats in the grocery store is still a ways off, and they are currently very expensive to produce.[h]

---

[a] Kelsey Piper, "The Rise of Meatless Meat, Explained," Vox, May 28, 2019, https://www.vox.com/2019/5/28/18626859/meatless-meat-explained-vegan-impossible-burger.

[b] Joanna Blythman, "The Quorn Revolution: The Rise of Ultra-Processed Fake Meat," *Guardian*, February 12, 2018, https://www.theguardian.com/lifeandstyle/2018/feb/12/quorn-revolution-rise-ultra-processed-fake-meat.

[c] Blythman, "Quorn Revolution"; Adele Peters, "Here's How the Footprint of the Plant-Based Impossible Burger Compares to Beef," Fast Company, March 20, 2019, https://www.fastcompany.com/90322572/heres-how-the-footprint-of-the-plant-based-impossible-burger-compares-to-beef.

[d] "Plant-Based Market Overview," Good Food Institute, November 27, 2018, https://www.gfi.org/marketresearch.

[e] Piper, "Rise of Meatless Meat."

[f] Neil Stephens et al., "Bringing Cultured Meat to Market: Technical, Socio-political, and Regulatory Challenges in Cellular Agriculture," *Trends in Food Science & Technology* 78 (August 1, 2018):160, https://doi.org/10.1016/j.tifs.2018.04.010.

[g] Hanna L. Tuomisto and M. Joost Teixeira de Mattos, "Environmental Impacts of Cultured Meat Production," abstract, *Environmental Science & Technology* 45, no. 14 (July 15, 2011):6117, https://doi.org/10.1021/es200130u.

[h] Jacob Bunge, "Cargill Invests in Startup That Grows 'Clean Meat' from Cells," *Wall Street Journal*, August 23, 2017, https://www.wsj.com/articles/cargill-backs-cell-culture-meat-1503486002.

What Is Being Done for Beef Cattle?

Researchers are examining how to adapt livestock production to current and future conditions. For instance, a University of Florida group is searching for the gene that regulates body temperature in the hopes of breeding beef cattle less sensitive to higher temperatures.[27]

Many producers provide shade in pastures and feedlots and use sprinkler and evaporation systems to cool cattle. Studies have also shown that in hot weather, cattle fed high-roughage, low-energy diets had lower body temperatures.[28] These adaptations minimize the amount of stress on cattle in an increasingly warm climate.

Producers can also reduce methane emissions from cattle. One dietary option is to feed them food waste, and there is plenty available, including wheat by-products, distillers' grains, and cannery and bakery wastes. Such an approach helps address waste management and pollution and the production of additional greenhouse gas emissions, such as would be created when wastes are put into landfills.[29]

Scientists now also see potential to modify the gut microbes in ruminants so they'll produce fewer greenhouse gases.[30] Adding small amounts of kelp to dairy cows' diets can reduce methane emissions and should work with beef cattle too.[31] Since all beef spend some portion of their lives grazing, improvements in the nutritional quality of forages would also help. The higher the forage quality, the more rapid the growth of the animals and the less methane produced over their lifetimes. Scientists in New Zealand demonstrated that dairy cattle fed on a forage called birdsfoot trefoil released about one-third less methane.[32] Other options to reduce methane production include feeding high-quality forages such as alfalfa when cattle are not on pasture (for example, during the winter), grinding the forage to improve digestibility, and fermenting the forage to produce silage.

Poultry

Roasted, fried, tossed in barbeque sauce, thrown on a pizza, or added to a salad, chicken is popular around the world. In 1928, the Republican Party promised a "chicken in every pot" as part of Herbert Hoover's presidential campaign.[33] In the early 1900s most farmers kept chickens as a small side business and a source of eggs for their own consumption. In 1923, Cecile Steele of Delaware mistakenly received 450 more chickens than she ordered and decided to raise them to sell as meat birds. She sold the chicken for five dollars per pound, an exceptionally high price at the time. Her business rapidly grew, and three years later she had 10,000 chickens, all in a barn. Her innovative shift to indoor production made chickens reliant on grain feed as opposed to foraging outdoors.

In the early 1940s a scientist named Thomas Jukes discovered that chickens given antibiotics resisted the diseases that spread when they were raised in confined quarters, plus they gained 20% to 25% more weight and required less food than other chickens. Antibiotics cost chicken farmers less than a penny per pound of animal, resulting in increased profit.

At this time, however, chickens raised for meat were quite different than the chickens we are familiar with today. They were mostly dark meat and rather tough, requiring a significant amount of cooking time. To make chicken more appealing, the Great Atlantic & Pacific Tea Company (A&P) partnered with the USDA and launched a nationwide contest in 1945 called the "Chicken of Tomorrow." The hope was to develop a breed that grew faster and bigger and gained weight where needed most.[34]

The chickens developed for this contest established the bloodlines for today's chicken, though chickens now produce about twice the meat in half the time it took in the 1940s. Chicken has become an affordable form of protein considered healthier than beef or pork because it has fewer calories and lower fat content.

Consequently, chicken is now the most widely consumed meat in the world. Global per capita consumption is 31 lb. (14 kg) per year, with US consumers eating about three times that amount.[35] In 2018, about 106 million tons (96 million metric tons) of broiler meat were produced worldwide, with the United States and Brazil being the largest producers.[36]

People also enjoy chicken eggs—scrambled, sunny side up, over easy, or in many dishes—and they are a good source of protein, minerals, and vitamins. China, the world's leader, produced 460 billion eggs in 2018, followed by the United States, at 109 billion.[37] In the United States, the state of Iowa leads egg production with 57 million laying hens in 2018.[38]

What Is Changing for Chickens?

Heat stress is the main concern for laying hens and chickens raised for meat. Researchers who housed layers (hens) under very warm conditions observed that egg weight, density, and shell thickness were all negatively affected.[39] The US poultry industry loses about $128 million each year owing to heat stress.[40] In 2012, a heat wave killed an estimated five hundred thousand laying hens housed in uncooled sheds in Brazil, and such events are expected to increase in coming years.[41]

Because most chickens are raised indoors in the United States, the temperature can generally be regulated and heat stressed minimized. In addition, the artificial lighting cycles within the houses can be manipulated (along with the amount of food) to increase egg size and production. In the United States and other developed countries, producers can build and maintain specialized housing. But in hot and humid regions where sophisticated housing is not an option, heat stress is becoming a major constraint to the poultry industry.[42]

Producers also face an increasing risk of extreme weather events—intensified by climate change—directly damaging poultry facilities. For example, Hurricane Florence in 2018 damaged North Carolina's agricultural industry, causing the loss of 3.4 million chickens.[43]

What Is Being Done to Protect Chickens?

Producers keep chickens comfortable in warm weather by means of evaporative cooling systems, which can be expensive. Some systems draw air through large porous pads and blow the cooled air through the chicken house. Fogging and misting systems spray fine droplets that cool the air when they evaporate. As the climate warms, some chicken-farming operations might move north and to cooler climes.[44]

Certain breeds of chicken are more tolerant to higher temperatures than others, and chicken producers are seeking them. Breeders are working to incorporate the "naked neck" trait into commercial lines, resulting in less feather growth.[45] Less feather coverage helps keep the animal cool.

In addition to adapting to climate change, some producers are exploring ways to mitigate their greenhouse gas emissions. Waste products from chicken farming can provide a nutrient-rich fertilizer for crops or be processed in an anaerobic digester to produce biofuel for generating electricity. Energy efficiency along the chicken meat supply chain can help operators of farms, slaughterhouses, processing plants, storage facilities, vehicles, and refrigeration units cut energy costs and reduce emissions.[46]

Fish and Other Aquatic Foods

In a shared fish, there are no bones.

DEMOCRITUS OF ABDERA

The aquatic world—rivers, lakes, and oceans—provides a wide variety of offerings. Globally, hundreds of different kinds of fish and shellfish are available for consumption, but people in the United States eat only a few types. The most commonly consumed are shrimp, which is typically imported from China or Thailand, salmon, and canned tuna. Tilapia, Alaska pollock, pangasius (a fish from Vietnam), cod, crab, catfish, and clams complete the list. The oceans provide vegetables too, such as kelp and sea beans.

Freshwater and oceanic fish became part of the human diet approximately forty thousand years ago.[47] Today, 55% of the world's oceans are fished, an area more than four times that of the world's agriculture.[48] Most commercial fishers stay within a country's Exclusive Economic Zone, which extends no more than 200 nautical miles (370 km) off a country's coastline, though some fish on the high seas. Commercial fishers catch 80% of their fish with various nets, the rest with fishing lines that are either long or much shorter and attached to a pole. Shellfish are harvested by dredging, traps, diving, or by hand.[49]

Fish are one of the most widely eaten foods in the world, and their popularity is expected to increase. In 2017, the global catch of oceanic fish was estimated to be 193 million tons (175 million metric tons) and valued at $125 billion. By far, China catches the highest volume of fish globally and also has the largest seafood- and fish-processing industries.[50] In 2017, the US fish industry harvested nearly 5 million tons (4.5 million metric tons) of fish and shellfish, valued at $5 billion. The country also imported 3 million tons (2.7 million metric tons) of seafood. Unfortunately, and despite a 45% increase in the number of fishing vessels globally from 1950 to 2015, both the global catch per amount of effort and fish stocks have been steadily declining.[51]

A rapidly increasing global population needs nutrient-rich fish, and aquaculture—the farming of aquatic animals and plants in freshwater and seawater—is more than filling the gap. According to the FAO, aquaculture

yielded 88 million tons (80 million metric tons) in 2016, over 45% of the world's fish supply. Aquaculture has helped to increase per capita fish consumption worldwide, which is estimated to be about 44 pounds (20 kg) per person annually.[52]

People in the United States feast on much less fish than the global average, but consumption is increasing, possibly because the country's dietary guidelines suggest eating 8 oz. (227 g) per week. Fish are a major source of omega-3 fats. They are rich in vitamins and other nutrients, low in saturated fat, and they help reduce the risk of dying from heart disease.[53]

What Is Changing for Ocean and Freshwater Life?

The vast oceans, which cover over 70% of the earth's surface, are acidifying and warming, as mentioned earlier. These major changes alter the growth, reproduction, survival, and interactions between species. And changes in the climate intensify existing stressors on oceans, such as overfishing, pollution, disease, and invasive species.

Most ocean plants and animals are extremely sensitive to temperature. To illustrate, as previously mentioned, phytoplankton populations have declined 20% in some regions over recent decades as the oceans have warmed.[54] Krill, tiny crustaceans that look like shrimp, feed on phytoplankton and are particularly common in the Antarctic and Arctic Oceans, though their distributions are shifting as the water warms.[55] Both the decline in phytoplankton—the basis of the food chain—and changes in krill distribution have profound implications for life in the oceans.

Increasing water temperatures are also causing mussels to lose their ability to form strong anchor points, making them more susceptible to being dislodged by waves and dying. The destruction of mussel beds not only dashes hopes for delicious dishes such as moules à la marinière but also destroys critical habitats for other marine creatures. In 2019, mussel populations along the Northern California coast died because of higher-than-normal temperatures combined with low tides and few afternoon breezes to help with cooling. Scientists suggest that climate change led to the die-off.[56]

In the Bering Sea, one of the most productive fishing regions in the world, warming conditions are causing the sea ice to retreat. Over 40%

of the US annual fish catch is from the Bering Sea—mainly pollock used in fish products from fish sticks to imitation crab. Retreating Arctic sea ice threatens the food supply of young pollock, which feed on the algae that grows under the ice. As the sea ice retreats, young pollock have less algae to feed on, their cold habitat is smaller, and they are preyed upon more by older, cannibalistic pollock. These factors threaten the overall stock of pollock, which is worth over $1 billion annually.[57]

Rising ocean temperatures deplete the oxygen content so essential for many forms of marine life, including fish. Scientists have recorded oxygen declines over large ocean and coastal regions and expect the phenomenon to expand as the oceans continue to warm.[58] Warming is also likely responsible for a massive starfish die-off along the west coast of North America, which may have a cascading effect on many aspects of this vast marine ecosystem, including the survival of abalone.[59]

Increasing temperatures also result in fish and lobster populations moving to cooler waters, and if they migrate across political boundaries, disputes can ensue over who has the right to catch them. By shifting their territories, Atlantic mackerel, Alaskan salmon, and New England lobsters have already triggered such battles, which are expected to increase as fish populations continue to relocate.[60]

Conflicts occur within domestic territories as well. For example, in 2016 black sea bass and summer flounder, typically found off the southeastern states, were popping up near Massachusetts. Fishing regulations limited the catch of northeastern fishing vessels but allowed fishing boats from North Carolina to catch their legal quotas—about ten times more than locals.[61]

As ocean acidity and temperatures increase, coral reefs are declining, and the habitats of many types of fish and other critically important species are being lost. Scientists have determined that acidic conditions interfere with the shell formation of young oysters in hatcheries and elsewhere. This finding has clear implications for the oyster farming industry, which is worth $273 million along the West Coast alone. Hatcheries have ameliorated the problem by manipulating water chemistry and continue to supply West Coast farms, but many see the threat as another canary in the coal mine for marine life.[62] Finally, when fish are exposed to the higher carbon dioxide levels anticipated by the end of the century, their sense of smell can be impaired, limiting their ability to detect food, predators, and safe habitats.[63]

A warming gulf of Maine

Off the northeastern seaboard, the Gulf of Maine spans an area larger than Maine itself. In this productive fishery, cold water from the north meets warm water from the south, and they circulate counterclockwise, stirring nutrients and oxygen so marine life can thrive. Normally surface waters in this area are cold: 50°F (10°C) in the winter and 63°F (17°C) in the summer. But meltwater from Greenland's glaciers is disrupting ocean circulation and pushing warm water into the Gulf. In the past thirty years, it has warmed faster than 99% of the world's oceans, and between 2003 and 2018, it warmed at more than seven times the average global rate, affecting fishing patterns and productivity.[a]

---

[a] Michael Carlowicz, "Watery Heatwave Cooks the Gulf of Maine," Global Climate Change: Vital Signs of the Planet, September 12, 2018, https://climate.nasa.gov/news/2798 /watery-heatwave-cooks-the-gulf-of-maine.

While changes in the ocean threaten some species of fish and shellfish, other organisms—such as pathogens that cause disease in marine animals and humans—are shifting in distribution or thriving. Dermo disease, a plague for oyster populations, has spread northward along the North American east coast thanks to climate change.[64] Warmer temperatures and increased salinity have apparently also increased the risk of *Vibrio* infections in humans. From 1996 to 2005 infection rates rose 41% in people who handled or ate shellfish. This pathogen, which is associated with shellfish in the southeastern United States, is highly resistant to antibiotic treatment, making it a serious threat to not just the livelihoods of many people but also their lives.[65]

Changes are also affecting the species of fish that live in our inland, freshwater lakes, rivers, and streams. Their waters too are warming, stream flows are changing, species are shifting, and less lake ice means less ice fishing. Favorites among fishing enthusiasts, such as trout, are struggling to survive in warming waters that contain less dissolved oxygen. Decreased snowpack in the winter results in lower stream flow in the spring and summer, limiting areas of acceptable fish habitat. And the parasitic sea lamprey, which benefits from the warming of the Great Lakes, has been

A lobsterwoman's game plan: Diversify

At 4:00 a.m. Krista Tripp wakes up, and by 5:30 she has loaded gear onto her boat in South Thomaston, Maine. Then, with her husband as sternman, she pilots the *Shearwater* into open water and begins fishing some of her six hundred lobster traps.

Tripp fished throughout high school and college, then with her dad, grandfather, and others for twelve years as she waited for her lobster license. Since 2015, when she became one of two hundred licensed commercial lobsterwomen in Maine, she's seen many changes.

"People need to go farther out to catch lobster than they used to," says Tripp. "When I was a kid there was just an incredible amount of lobsters inshore, and now they're not coming inshore as much as they used to. That's definitely a major concern." She says some lobster harvesters with federal licenses travel 20, 30, even 50 mi. offshore, depending on how big their boat is and how much fuel it can hold.

Tripp has heard forecasts of lobster starting to move into Canada if the water continues to get warmer. That's why on a typical day she comes home at 2:00 pm for a quick cup of coffee, and if the tide is in her favor, navigates to her oyster farm. "I started this," she says, "because of the climate change stuff going on and the different findings with what lobsters are predicted to do in the future. I'm just trying to diversify so I can be prepared in case something drastic happens."

Tripp bought Aphrodite Oysters in 2018 and farms about fifteen lines in the Weskeag River. On each floating line hang ten to twenty mesh bags

increasing in number and physical size, putting species such as lake trout, walleye, sturgeon, and catfish at increasing risk.[66] Finally, lake trout are feeding less in the warming nearshore waters and more in deeper cooler waters, potentially affecting the entire food web.[67]

What Is Being Done to Help Seafood Thrive?

Considering the scale of the oceans, the current trajectory of increased warming, acidification, and sea level rise, the challenges that climate change poses to our seafood supply might seem nearly impossible to address. And it's not just climate change impairing global fisheries. Scientists reported

with American oysters inside, where they are protected from green crabs and other predators. She grows the seeds (immature oysters) in one spot of this intertidal river, the bigger ones in another. In a third place she allows mature oysters to 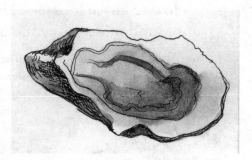 purge themselves of sand and other impurities before testing them to be sure they're safe for sale.

Tripp sells to local restaurants, seafood markets, and dealers. She says customers think the smaller oysters taste better. "That's really great, because it's a faster turnaround." Maine's water is usually so cold that larger oysters need three years to size up. "I do know that the waters are warming, and that might impact the farm in a positive way," she says.

How does she manage two businesses? "I'm pretty busy," says Tripp. Sometimes she gets home at 7:00 pm. "It's really hard doing both." But she's concerned about the outlook for lobster. "I don't think it's too much of worry in the near future, but I do know that it's potentially going to be a big worry in the next forty years or so. It might not be in my time, but if I do have children, I would like to leave something behind for them, some sort of business, and I just don't know if lobstering is going to be it."

that in 2015, 33% of the world's marine fish supply was overfished, and an additional 60% was fished at the maximum sustainable levels. What can be done? The solutions are mainly social and political.[68]

First and foremost, we need to reduce global greenhouse gas emissions to mitigate these negative changes. Harvesters can decrease emissions from marine fishing, which produces about 4% of the total food production emissions. Fishing pelagic species (that swim close to the ocean surface), for example, typically has a lower carbon footprint because the fish-to-fuel ratio is higher than with catching benthic, or bottom-dwelling, species.[69]

Experts also agree that to maintain an adequate supply of fish for an increasing global population, we need to manage wild fisheries wisely,

How to catch lobsters: Live in Maine and throw out your calendar

"The Casco Bay region is not the same as it was when I was a kid," says Steve Train, a full-time fisherman who sails the *Wild Irish Rose* out of Long Island, Maine. "Our bottom temperatures are warmer than they used to be."

Train, who has been fishing since he was eight, has harvested groundfish, lobster, scallops, shrimp, and sea urchins. He started setting lobster gear on his own boat in 1990 and now has eight hundred traps in island-studded Casco Bay. Train says almost every dime he has earned has come out of the ocean. Since 2012, he's been a member of the Atlantic States Marine Fisheries Commission, which manages the American lobster and several other species.

"Now it looks like we're at all-time highs with lobster," says Train, referring to Maine's midcoast region. Not so for lobstermen and women in Massachusetts and Rhode Island, where populations have plummeted. For more than fifteen years, rising temperatures in Gulf of Maine bottom waters have hastened the rate at which a lobster will grow, shed its outgrown shell, and harden its new shell. "We are seeing earlier molts in deeper water, further offshore—stuff we didn't see in the past," Train says. "I think a lot of that is because of the temperature at depth."

Train isn't worried the lobster fishery will collapse, but he has seen surprising changes in fish species. "For the last few years, the kids have been

reduce overfishing, and bolster aquaculture production. Such changes need to occur at the local, national, and global levels.[70] Some researchers suggest that despite the negative effects of climate change, improved fishery management could lead to higher yields and profits in the future.[71]

Aquaculture clearly has potential, but minimizing environmental damage is essential. Some aquaculture operations convert essential natural able to go down to the dock and jig squid at will. So that to me is an indication that our current climate isn't what our traditional climate was." He's also seeing unusual numbers of red hake and black sea bass. "Whether they're filling a spot vacated by another species that doesn't want to be there, or whether they're attracted because of the water temperature, I don't know."

Train acknowledges that his snapshot of the fishery is limited to his lifetime. "Do I personally think we have a problem with climate? Yes. Maybe it will flip back around, but it's not one of those things I like gambling with."

When he first started lobstering, Train recorded dates and locations in a little log. "In the last six to eight years, I've thrown the books out. I'm just trying to figure out where I should be next by what's going on now, not what used to happen. So I'm constantly adjusting and I'm cautiously fearful.

"For a harvester, the easiest way to recognize whether a fishery is healthy is the catch per unit effort, whether it's traps or trawls. So if your catch per trap is doing well, in your mind, the fishery is healthy. What we don't always take into account is that we are very adaptive in the way we fish. We congregate gear in an effective area. In a place that is doing well, we'll put more traps, so that our catch per unit effort will stay acceptable to us because we can't afford it not to be. It all comes down to sustainability, and we need to sustain the resource, but our first reaction is to sustain ourselves. We don't always see that we've got a problem coming."

Train says the effect of the acidifying oceans on lobster is still unknown. "The question becomes: Can the females that are egging out survive there? Can the larval lobster survive there? How long can we continue to harvest at a sustainable level? That has a lot to do with the water temperature and the acidification of the water. Some people will never believe it because they're still catching lobsters there, so we're not all looking at a big picture all the time."

wetlands into fish farms and feed wild fish to farmed fish, both questionable practices. Others contribute to water pollution or allow farmed fish to escape and breed with wild fish, altering wild populations.

To expand aquaculture sustainably the FAO recommends maintaining fish health, increasing water efficiency and feeding efficiency, and developing fish that are better adapted to wider salinity ranges.[72] Some countries have adopted a few of these recommendations, but wider acceptance could ensure successful and productive aquaculture for the future.

To keep up with the United States' growing demand for shrimp, producers in Asia have farmed intensively, degrading the vast coast-protecting mangroves, polluting local water supplies, and threatening species of native fish and plants. Certain management strategies can reduce these negative impacts. For example, shrimp and fish farms can be relocated away from mangrove forests or other natural habitats that require protection. To avoid contaminating local water supplies, fish farmers can limit the use of fertilizers and chemicals that kill pathogens. Last, shrimp farming can be integrated into mangrove cultivation, whereby flooded areas sustain shrimp production and beneficial mangroves simultaneously.[73]

Researchers suggest that genetically engineered fish will also have an important role in maintaining supplies. For example, genetically engineered salmon that grow faster and reach maturity in half the time of their nonengineered counterparts have been sold in Canada and may appear in US markets in the future. The fish are reared in tanks and so are not exposed to the diseases and parasites that sometimes afflict salmon raised in sea cages.[74]

Another option may be to shift what species we consume. For example, squid, octopus, and cuttlefish—which are booming in the oceans— are garnering attention from chefs around the world. Scientists suggest that these animals are benefiting from the changing ocean.[75] They adapt quickly because they grow rapidly and have fairly short life spans. As other seafood becomes more expensive, or supplies decrease, we may find these cephalopods on the menu more and more.

Summing It Up

As with most of the menu, the many choices for the main course are sparring with a changing climate. Increasing heat stresses animals. Seafood is strained by pollution, overfishing, and both the warming waters and acidification caused by climate change. At the same time scientists, farmers, ranchers, and fishers are working to keep these incredibly important foods on the menu. New breeds of chicken and cattle that are more heat tolerant are being developed along with improved cooling tactics and diets to reduce heat stress. Along with this progress, improved pasture management

Think creatively about local fish

Kyle Foley could tell sad stories about economic hardships from warming sea temperatures, like the closing of the northern shrimp fishery in 2013. But she'd rather focus on what the Gulf of Maine Research Institute (GMRI) does best: finding solutions.

About seventy staff members at this Portland, Maine, nonprofit study marine issues, educate students, and help harvesters and others to succeed. Foley manages the institute's Sustainable Seafood Program, which melds ecology with economics. Central to her work is the *Gulf of Maine Responsibly Harvested* eco-label. "Our program resonates for a lot of people," says Foley, "because it has both the place-based nature—identifying a product that's from this region—and the responsibly harvested verification."

Eight fish processors are partners, marketing the certified seafood to buyers that include three large food service companies. One of those companies, Sodexo, buys 82% of its white fish from the Gulf of Maine for eleven Maine school campuses, and staff are on track to reach 100%. As food service companies introduce students to a range of seafood, they're helping drive demand for underutilized species and educating a new generation of consumers.

In 2018, the Sustainable Seafood Program engaged forty-eight universities, schools, restaurants, and hospitals. "We've built up a lot of awareness with our restaurant partners, with retailers like Hannaford,

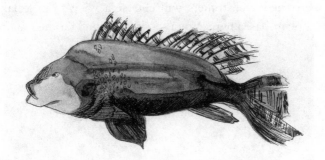

Black sea bass

(*Continued on next page*)

(*Continued from previous page*)

and with institutional dining services around these other species that are in the region, like pollock, redfish, hake, dogfish, and mackerel," says Foley. That awareness will be more and more important as species distribution continues to change over time. She predicts that black sea bass and squid, for example, will become important local fisheries.

The GMRI has also brought together more than four hundred people so they can educate each other in workshops. Says Foley, "We try to get fishermen, processors, distributors, chefs, retailers, anybody in the supply chain buying seafood, all in a room together for a day." Through conversation and informal sessions they share what's happening, like surpluses or demand for specific fish. Foley hopes the workshops will become a forum for talking more about how climate change is affecting the supply chain. "Whether you're here in the Gulf of Maine, or in another part of the country, or the world, we're going to have to think a little bit more flexibly about eating what fishermen can go out and catch."

helps to sequester carbon, and changes in diets reduce emissions from cattle. The oceans, a difficult challenge given their scale and the fundamental changes underway, call for improved global management along with an overall reduction in greenhouse gas emissions.

Maybe someday customers will choose a wing from a naked-neck chicken, a plant-based burger, or a serving of squid for their main course.

Grains, Starches, and Other Sides

Our main dishes often come with a side dish or two, maybe a potato, some creamy polenta, or another grain, such as quinoa, faro, or barley. This chapter takes a closer look at some of the grains, starches, and vegetables we put on our tables every day, or even at every meal.

Cereal grains, the most widely grown crops around the world, are planted on 1.8 billion acres (730 million hectares).[1] Over time, cereal grain production has become more efficient; globally, farmers now produce three times more per unit of land than they did in 1961.[2] Grains, including wheat, rice, maize, sorghum, and millet, provide the world with most of the nutrients and calories we need: 60% of total calories for people in many developing countries and as much as 80% in the poorest nations. In the United States, that percentage is much lower, and many grains are used instead for animal feed, bioethanol, sweeteners, or other purposes. For example, in the United States about three times as much corn goes into the production of sweeteners and starches than individuals directly consume.[3]

Environment, culture, and economics determine the types of grains grown around the world, with temperature and water availability the most significant factors in cereal grain growth. Varieties of wheat can thrive under a range of temperatures and are therefore produced widely across the globe, while sorghum and millet, more drought tolerant, tend to be grown in semi-arid climates, such as in India and African countries. Barley does well in cooler temperatures and is produced largely in the northern United States, Canada, and northern Europe.

Rice

With coarse rice to eat, with water to drink,
and my bent arm for a pillow—
I have still joy in the midst of all these things.

CONFUCIUS, TRANSLATED BY JAMES LEGGE

People eat rice in countless ways around the globe. In China, it arrives at the table steaming in a small bowl. A banana leaf enfolds it in many South Indian meals. In Trinidad, cooks simmer rice in coconut water until starchy, then steam it with cilantro-like shado beni, bird's-eye pepper, and garlic. In Iran, sadri rice comes on great copper platters studded with golden raisins, barberries, flowers, and blanched almonds, layered with saffron, and swirled with cardamom. In Japan, the United States, and Europe, deft hands wrap rice and fish into sushi. In addition to being eaten as a grain, rice also appears in the form of noodles, edible rice paper, rice cakes, dumplings, rice oil, vinegar, and alcoholic beverages.

Nutty basmati rice comes from the Himalayan foothills; arborio from Northern Italy; aromatic jasmine rice from the paddies of Thailand; and Camargue red rice from the marshes of Southwest France. There's short grain, long grain, white, brown, red, black, and wild. Rice is a ubiquitous staple, grown in over 113 countries, and this essential grain has nourished cuisines, cultures, and societies for thousands of years.

Rice is the most important food for over three billion people worldwide and provides more than 50% of the dietary calories for 520 million people in Asia.[4] It also is the primary source of employment and income for millions of households in developing countries and is critical to alleviating

global poverty and malnutrition. In Africa, rice is one of the most valued crops. It is considered so important that the United Nations designated 2004 as the International Year of Rice, the first year ever dedicated to a commodity.

Where and How Is Rice Grown?

While the exact geographic origins of Asian rice remain uncertain, experts agree that a wild species was first intentionally planted in India or China some eight to nine thousand years ago.[5] In Africa, a different species of rice was independently domesticated from its wild ancestor.[6] There is also wild rice, technically not a rice species, which indigenous North American cultures have considered sacred for thousands of years. Today California and Minnesota lead the way in US production of cultivated wild rice.[7]

China and India produce about half of the globe's milled rice, which totals roughly 547 million tons (496 million metric tons) annually.[8] The United States has some of the highest yields in the world and accounts for about 2% of global production, valued at about $2.8 billion in 2018.[9] Although the United States exports a considerable amount of rice, it also imports aromatic varieties such as jasmine and basmati.[10]

Rice is in the grass family and is typically grown in warm, wet conditions, often in standing water. The standing water in rice paddies controls weeds and makes many soil nutrients more readily available to the rice roots. It may also contribute to groundwater recharge. Rice is one of the few crops that can tolerate being submerged because it has special air channels that move oxygen from plant parts above water to those underwater. Farmers also grow rice without flooding, although doing so is more challenging and yields less than when rice is grown in paddies. Growing rice this way depends on annual rainfall or, in some instances, irrigation.

In Asia and to some extent in Africa, growing rice is labor intensive, starting when farmers flood the paddies and work the soil to prepare it for transplanting seedlings. Smallholder farmers will hand weed the crop multiple times during the growing season, which can be up to four months. They often harvest it by hand using sickles and knives. Mechanization of some of these practices is increasing but often is cost prohibitive for small

operations. In contrast, rice production practices in the United States, Europe, Australia, and South America are much more mechanized, including the aerial application of seed, fertilizer, and pesticides.

After harvest, rice is dried and milled to remove the husk, leaving brown rice. Additional processing removes the bran layer, which is rich in oils and micronutrients. The result is white rice. The bran is used in making breakfast cereals, for baking, or as a highly valued enrichment for animal feed. In the United States, rice mills use laser sorters to remove broken or discolored kernels, resulting in uniformly high-quality rice.

What Is Changing for Rice?

The United Nations' Food and Agriculture Organization estimates that unless agricultural practices improve significantly, water scarcity and increased temperatures may cause rice yields in India, Indonesia, the Philippines, Thailand, and Vietnam to drop 50% by 2100.[11] Vulnerability to climate change in these drought-prone regions is extremely high, and the

effect on food security will be immense. In coastal regions, saltwater intrusion as sea levels rise increasingly threatens rice production. Scientists estimate that in some coastal areas of Bangladesh, rice yields could decline by 15% by 2050 because of increases in soil salinity.[12]

The increase in occurrence of droughts also affects rice production in the United States. The situation is especially acute in California, where severe droughts result in water shortages. During California's 2011–2014 drought, thousands of acres normally in rice production were left fallow.[13] Risks to California's rice production should get the attention of US sushi restaurants, which obtain most of their rice from California.

Without improved varieties or effective adaptation, some scientists predict rice yields will decrease 3.2% for every 1.8°F (1.0°C) increase in temperature.[14] Spikes in temperature can cause plant sterility if they occur during the flowering stage, and other studies show that high nighttime temperatures reduce grain yields.[15] It's not just temperatures that are affecting rice. Researchers have demonstrated that future levels of carbon dioxide will diminish the vitamins and minerals in rice, which will likely affect the health of millions of poor in the major rice-consuming countries.[16]

Rice farming also contributes significantly to climate change. The microbes in rice paddies' anaerobic soils decompose organic matter and produce about 9% of global anthropogenic methane emissions.[17]

Cutting methane from rice paddies

In some places, rice paddies are flooded not only during the growing season but also off-season to provide habitat for waterfowl and other wildlife. "As organic matter decomposes in the anaerobic environment under the flooded part of field, it produces methane," says Allison Thomson, science and research director at the nonprofit Field to Market, which coordinates farmers, businesses, and other partners in reaching sustainability goals. To save water and slash the amount of methane produced, scientists are studying how to cut the time a field is flooded.

Thomson thinks a warming climate might bring opportunities to rice farmers. "There's a way to grow rice called ratooning," she explains. "You grow a crop, then you harvest and let the rice grow back. It's effective, but it's only possible where you have lots of warmth." Thus ratooning is limited to the far south right now, but researchers are experimenting with it farther north.

What Is Being Done for Rice?

Globally and in the United States, researchers are breeding rice varieties for resilience to climate-change-induced stresses, such as drought, salinity, and high temperatures. The International Rice Research Institute, headquartered in the Philippines, is recognized globally for its role in improving rice varieties.[18] In the United States, a joint team from the University of Nebraska–Lincoln, Arkansas State University, and Kansas State University focuses on improving wheat and rice yields in stressful environments.[19] And scientists at the Rice Experiment Station, funded primarily by California rice growers, are developing improved varieties to maintain high yields and quality.[20]

Crop management improvements can also help alleviate the new climate stresses. For example, smallholder farmers using SRI—the System of Rice Intensification—can improve soil health and get by with fewer plants and less water. Demonstrated in over fifty countries, SRI helps farmers with limited resources increase their rice yields and resilience.[21]

Farmers in the United States and around the world are also practicing alternate wetting and drying (AWD). With AWD, farmers don't flood their fields all season long but vary the water level, even letting the field surface dry out before refilling. Alternate wetting and drying reduces overall water consumption, and dry fields interrupt anaerobic processes and reduce methane emissions.[22] Scientists have recently reported that incorporating rice straw over several years may also reduce methane emissions from rice paddies.[23] In 2018, Microsoft purchased carbon credits (offsets for its own carbon emissions) through the American Carbon Registry from a group of rice farmers in Mississippi and California who were practicing AWD.[24] While the payouts were relatively small, this program might offer a new approach for reducing agricultural emissions.

Additionally, farmers improve water use by carefully managing water flow from paddy to paddy, leveling the land where feasible to promote uniform water distribution and installing water recovery systems to retain excess irrigation water and rainfall runoff in reservoirs. The latter tactic can lessen the need to pump water from the ground. Experts are also investigating infiltration basins that channel water back into local groundwater systems.[25]

How can farmers increase rice production even as climate change makes it harder worldwide? One answer is multinational cooperation. Producers need more resilient varieties and growing practices that mitigate their own impacts on the climate. For millions of people on the planet, rice is life, and it is increasingly at risk.

Wheat

Wisdom, Power, and Goodness meet
In the bounteous Field of Wheat!

HANNAH FLAGG GOULD, "THE WHEAT FIELD"

If rice is the most important grain in the world, wheat is the second.[26] People eat wheat in nearly every country: *chapati* and *naan* in India, pot stickers in Japan, *couscous* in Northern Africa, *tabbouleh* in the Middle East, and *alfajores* for dessert in Peru. Wheat provides a quick and easy bowl of cereal in the morning. Rich in dietary fiber, it feeds people, livestock, and fish. Wheat flour goes into bread, cakes, cookies, pies, pastries, crackers, pasta, and flour tortillas, as well as beer, whiskey, and some vodkas. It also fuels cars and is used in making textiles and plastics.[27] With

enough water and good pest management, this desirable, highly versatile crop yields well.

Archaeological sites in Israel show that humans gathered wild wheat twenty-three thousand years ago. As people began to transition from hunter-gatherer societies to agricultural ones, they chose to plant wild wheat varieties that could produce high yields. Modern wheat originated about ten thousand years ago in the Karacadag mountains in what is today southeast Turkey. Wheat is considered one of the eight founder crops on which all agriculture is based. Across time and around the globe, humans have grown nearly twenty-five thousand varieties.[28]

Where Is Wheat Grown?

Both the United States and China are major wheat producers; however, the crop is grown all over the world. In many developing countries, small-

The eight founder crops

According to some archaeologists, the eight founder crops include cereals and other crops that originated in the Fertile Crescent region, which today consists of southern Syria, Jordan, Israel, Palestine, Turkey, and Iran's Zagros foothills. The cereals are thought to be einkorn wheat, emmer wheat, and barley, and the other crops are lentils, peas, chickpeas, bitter vetch (related to fava or broad beans), and flax (which produces fiber for linen and linseed for oil). Debate continues over the exact plants and the number of plants that people grew thousands of years ago.[a]

---

[a] K. Kris Hirst, "The Eight Founder Crops and the Origins of Agriculture: Re-imagining the Beginning of Farming," ThoughtCo, updated August 31, 2018, https://www.thought co.com/founder-crops-origins-of-agriculture-171203.

holder farmers grow wheat for their own consumption. In the United States, forty-two states produce six different classes of wheat: hard red winter, hard red spring, soft red winter, soft white, hard white, and durum. Each goes into different products, from pasta to bread rolls to Asian-style noodles. Winter wheat requires vernalization (exposure to cold temperatures for multiple weeks at the seedling stage), so is planted in the fall and harvested in the spring or summer. Spring wheat, which does not require vernalization, can be planted in the spring and harvested in the summer or planted as an overwinter crop in milder climates.

What Is Changing for Wheat?

While wheat can withstand some variability in temperature, certain varieties are sensitive to excess precipitation and disease. Researchers predict wheat production will decrease by 6% for every increase in 1.8°F (1°C).[29] Scientists more recently reported that if we don't curb climate change, up to 60% of the current global wheat-growing area will likely be affected by simultaneous severe water scarcity by the end of the century.[30]

A 2019 report by the US Department of Agriculture's Economic Research Service suggests that overall the impacts on wheat in the United

Ancient grains and biodiversity

Consumers are now buying a rainbow of grains, such as quinoa, teff, millet, sorghum, and spelt.[a] These grains diversify the diet and help protect soils. Quinoa and other foods described as "ancient" have been on the rise for nearly a decade, and consumers perceive them as healthier options. A pseudo-cereal, quinoa is actually a seed crop that bears the tiny "fruit" with which many people are familiar. Grown mainly in Bolivia and Peru, quinoa can survive in a variety of climates and endure harsh conditions, including poor soil, drought, and saltwater intrusion. It is also very nutritious, with a high protein content and no gluten.[b]

[a] Nico Roesler, "The Future of Ancient Grains," *Food Business News*, April 19, 2018, https://www.foodbusinessnews.net/articles/11644-the-future-of-ancient-grains?v=preview.
[b] Karina B. Ruiz et al., "Quinoa Biodiversity and Sustainability for Food Security under Climate Change. A Review." *Agronomy for Sustainable Development* 34, no. 2 (April 1, 2014):349–59, https://doi.org/10.1007/s13593-013-0195-0.

States will be modest and variable as we approach the end of the century. For example, at higher latitudes, winter wheat will likely benefit from a changing climate in contrast to more southerly regions.[31] Warmer winter temperatures may also mean higher survival rates of insects and potentially larger populations during the growing season.[32]

Wheat is also susceptible to a number of fungal diseases, which may pose either more or less risk across the globe as the climate changes. For example, *Septoria tritici* blotch will likely decrease in France, while *Fusarium* head blight will likely increase in the United Kingdom in coming years.[33]

What Is Being Done about Grains?

Countries are looking into selective breeding and other adaptive measures to protect grain crops. For example, in China farmers are adjusting their cropping patterns (what is grown in a specific area in a given time period), planting seeds later in the year to avoid extreme heat and drought, and using water-saving techniques and irrigation.[34] Studies of wheat production in Australia show that by planting earlier in the season with a type of wheat that grows more slowly wheat farmers can adapt to a longer grow-ing season.[35]

The International Maize and Wheat Improvement Center, headquar-tered in Mexico, is working across fourteen countries to understand how to make both wheat and maize production more sustainable, as well as more tolerant to drought, heat, and disease. Researchers working with the Seeds of Discovery program are trying to identify the genes that will help wheat varieties survive in a changing climate.[36]

Potatoes

Let the sky rain potatoes.

SHAKESPEARE, *THE MERRY WIVES OF WINDSOR*

Last in the journey around side dishes comes the ubiquitous potato. Pota-toes are a favorite in the United States and beyond, and the delicious, plen-tiful tuber can be enjoyed mashed with butter or gravy, baked, stuffed, fried, roasted, riced, or boiled. Picture shepherd's pie in the United King-dom, Scotland's neeps and tatties, Indian curry dishes, poutine in Can-ada, or the US favorite, potato salad. The United States alone grows several hundred varieties, among them russet, long white, round white, purple, blue, yellow, and fingerlings. Potatoes are characterized by their skin and

flesh colors, firmness, texture, and flavor, and they come fresh, instant, or frozen as wedges, tots, hash browns, or crinkle-cut fries. When prepared correctly, one medium potato has half of the daily recommended vitamin C, 20% of the daily recommended potassium, and high levels of protein and carbohydrates.[37]

The versatility of spuds goes beyond just slicing or dicing. Potato flour binds meat mixes and thickens soups and gravy. Potato starch also thickens stews and sauces and binds biscuits, ice cream, and cake mixes. To top it off, taters go into alcoholic beverages such as akvavit and vodka and over 91,000 tn. (82,500 t) were used for livestock feed in 2017 in the United States.[38]

The potato belongs to the nightshade family, which also includes eggplant, peppers, and tomatoes, while the sweet potato belongs to the morning glory family. Potatoes were likely first cultivated in the Andes mountains seven thousand years ago, and eventually they arrived in Europe. At first, Europeans were reluctant to eat them, but because they were relatively easy to grow and cook, had a long shelf life, and were nutritious, they soon grew in popularity. Potatoes arrived in North America in the late 1600s, and the White House first served French fries during Thomas Jefferson's presidency.[39] Potato cultivation spread among developing countries in the 1960s, especially in China and India.

Where Are Potatoes Grown?

In the United States, the potato is the leading vegetable crop, and the average person eats about 115 lb. (52 kg) a year.[40] They are grown commercially in thirty states, but Idaho and Washington State together produce over half the crop. In 2017, potatoes were harvested from over one million acres (405,000 ha), yielding a total of 22 million tons (20 million metric tons) valued at $3.8 billion.[41] A lot of US potatoes and potato products are exported, including 100,000 tn. (91,000 t) of frozen fries. And the United States imports 180,000 tn. (163,000 t) of processed and fresh potatoes, mainly from Canada, each year.[42]

Globally, potatoes are a staple for 1.3 billion people.[43] China is the top producer, at 99 million tons (90 million metric tons) per year, India is second, and the United States is fifth.[44] Farmers in Bangladesh have found potatoes to be a valuable cash crop, while farmers in Southeast Asia are more focused on supplying food processing companies. Potatoes have also become a valuable staple crop in areas of Africa. The spread of the potato across the globe occurred so rapidly that the United Nations named 2008 the International Year of the Potato.[45]

What Is Changing for Potatoes?

With warmer temperatures and changes in water availability, the risks to potato production are increasing. Nighttime temperatures are particularly important for potatoes: if they rise above 77°F (25°C), tuber growth declines. This increase in temperature alone could undermine global potato production.[46]

In 2018 severe heat and drought hit spuds hard in England and Wales, cutting yields 20% and resulting in the fourth smallest harvest of potatoes since 1960. Individual potatoes were smaller, shortening British chips by 1 in. (3 cm).[47] Under a business-as-usual scenario in which greenhouse gases continue to increase, researchers estimate up to 95% of English and Welsh potato-growing land currently dependent on rainfall will become unsuitable for production by 2050 because of increasingly dry conditions.[48] With irrigation, other areas could remain productive but will be constrained by limited access to that water.[49]

In the United States, potato yields in eastern Washington State could drop as much as 22% by 2080 owing to rising temperatures.[50] Elsewhere, models suggest potato yields in sub-Saharan Africa could decline by as much as 50% by midcentury because of increasing temperatures. Coupled with increasing populations, the region could face serious food shortages if heat-resistant potato varieties are not developed and adopted.[51] Additionally, regions in India will likely see yield losses of 6% by 2050.[52] In general, increasing carbon dioxide levels will help offset the negative impacts of climate change until midcentury, but by the end of the century increasing temperatures will override this positive impact.[53]

What Is Being Done for Spuds?

As is true with so many of our favorite foods, keeping potatoes on the menu is going to take multiple approaches. Where practical, potato growers may have the option to move production to higher latitudes or to higher altitudes where temperatures may be more conducive to potato growth and development. Other options include using later-maturing varieties in temperate regions or earlier-maturing varieties in highland tropics to fit local climatic conditions.[54]

To maintain current production, experts must breed potato varieties resilient to the new conditions. In the United States, selecting for heat tolerance is a top priority, along with resistance to more pests and diseases. Globally, scientists also see the need for heat-tolerant cultivars along with varieties developed for drier climates. The genetic material for these breeding programs comes from wild and domesticated species with drought tolerance.[55]

The International Potato Center in Lima, Peru, has a collection of over seven thousand types of native, wild, and improved varieties.[56] Scientists can use the genetic diversity stored there to build potatoes with disease and insect resistance as well as improved yields, flavors, textures, colors, and many other characteristics necessary for resilience.

Breeding a better potato used to be a very slow and difficult process, but genetic engineering and hybrid breeding have expedited it.[57] Thanks to scientists who are breeding potatoes for resilience to climate change and farmers who are changing where and how they grow potatoes, the ubiquitous spud should be on the menu for the foreseeable future. Constant vigilance and support from governments, the private sector, and industry will make sure that stays true.

Summing It Up

As with nearly everything else on the menu, side dishes like grains and starches are changing. Several of the grains are globally important, planted to millions of acres, and the source of nutrition for billions of people. Potatoes are deeply imbedded in cultures and economies and served in innumerable ways. Increasing temperatures, droughts, and extreme weather are taking their toll worldwide, and the potential loss of nutritional value of these staples looms. In the future, your side dish might be pasta made from ancient grains, a legume-based bread, or a potato similar to those once eaten by the Incas.

Researchers, farmers, and others are battling this threat by developing improved crop varieties, implementing water conservation practices,

Genetic engineering—GMOs

People differ widely and often emotionally in their opinions of genetically modified or engineered organisms—often referred to as GMOs in the popular press—especially as genetic modification relates to health and environmental effects. Nonetheless, the National Academies of Sciences, Engineering, and Medicine concluded that consuming foods containing ingredients from genetically engineered crops poses no more risk than consuming the same foods using ingredients from crop plants developed with conventional plant-breeding techniques. The report also indicated that the use of GMO crops is generally profitable and poses few environmental problems. At the same time, the National Academies report stated that every genetically engineered crop should be evaluated on a case-by-case basis.[a]

Despite the evidence, about two-thirds of those surveyed in the United States oppose this technology, and of these 45% indicate that they are absolutely opposed to its use in food, regardless of the risk or benefits.[b] Yet in another survey, over half indicated that they knew very little or nothing at all about genetically modified foods, and one-quarter indicated that they had not heard of them.[c] The evidence supports the use of GMO crops, and they hold tremendous potential in saving the menu and our food supply. They are not a "silver bullet" but an important tool in the toolbox. (More detailed information about plant breeding and genetically engineered crops is provided in the "Solutions" section.)

---

[a] National Academies of Sciences, Engineering, and Medicine, *Genetically Engineered Crops: Experiences and Prospects* (Washington, DC: National Academies, 2016), 5–28, https://doi.org/10.17226/23395.

[b] Sydney E. Scott, Yoel Inbar, and Paul Rozin, "Evidence for Absolute Moral Opposition to Genetically Modified Food in the United States," abstract, *Perspectives on Psychological Science* (May 22, 2016):315, https://doi.org/10.1177/1745691615621275.

[c] W. Hallman, C. Cuite, and X. Morin, "Public Perceptions of Labelling Genetically Modified Food" (working paper, Rutgers School of Environmental and Biological Sciences, January 2013), 3, http://humeco.rutgers.edu/documents_pdf/news/gmlabelingperceptions.pdf.

Mixed veggie forecast

We can't forget how the vegetables we need and enjoy are changing. Two teams of scientists reviewed the knowledge about the effects of climate change on vegetables. Not surprisingly, they reported that increasing air temperature or cutting water availability in half reduced average yields more than 30%. On the other hand, a longer growing season could result in multiple crops per year and higher overall production. Unfortunately, some of our favorite vegetables, such as asparagus, need a cold season, and it's warming. The scientists recommend adaptation strategies such as development of climate-resilient varieties and more efficient use of water resources.[a]

Vegetables grown under different levels of carbon dioxide (simulating what the future holds) yield mixed results. For example, protein, iron, and zinc decline while concentrations of total flavonoids, calcium, glucose, and total antioxidant capacity increase. From a practical standpoint, the nutritional quality of vegetables could possibly be improved by selecting varieties that respond well to higher carbon dioxide levels.[b]

---

[a] Mehdi Benyoussef Bisbis, Nazim Gruda, and Michael Blanke, "Potential Impacts of Climate Change on Vegetable Production and Product Quality—a Review," *Journal of Cleaner Production* 170 (January 1, 2018):1602–s0, https://doi.org/10.1016/j.jclepro.2017.09.224; Pauline F. D. Scheelbeek et al., "Effect of Environmental Changes on Vegetable and Legume Yields and Nutritional Quality," *Proceedings of the National Academy of Sciences* 115, no. 26 (June 26, 2018):6804–9, https://doi.org/10.1073/pnas.1800442115.

[b] Jinlong Dong et al., "Effects of Elevated CO_2 on Nutritional Quality of Vegetables: A Review," *Frontiers in Plant Science* 9 (2018):1–11, https://doi.org/10.3389/fpls.2018.00924.

Tomato processor measures rain in feet

"My New Year's resolution was 'don't talk about 2018,'" says Ken Martin, director of agricultural operations for Furmano Foods. He's kidding—but only partly.

A family-owned business since 1921, Furmano Foods is the largest tomato processor and dried bean canner in the eastern United States. The Furmano brand name is known for about 150 different tomato products that they "dice, spice, and slice," says Martin, including whole peeled tomatoes and tomato sauce. Based in Northumberland, Pennsylvania, Furmano also processes three- and four-bean salads and edible dry beans, including red kidneys from New York, black beans from Minnesota, and chickpeas from California. Furmano wholesales these to Sysco, US Foods, prisons, hospitals, and grocery stores. The company grows 725 ac. (293 ha) of tomatoes on its own farm and contracts with twenty-two growers farming another thousand acres (400 ha) in Pennsylvania, New Jersey, Maryland, and Delaware.

"We're stretched out in the region to protect us from risk," says Martin. Nonetheless, in 2018 all their tomato farms were hit with heavy rain. "You know it's bad when you're measuring rain in feet, not inches. We farm river bottom, and the river came up three times in season." Soils were saturated, they couldn't get on the land, and fruit rotted. "We had eighty acres of

(Continued on next page)

(*Continued from previous page*)

tomatoes we never touched with a harvester." After Furmano lost 40% of its crop, it bought 9,000 tn. (8,200 t) of tomatoes from Ohio and paste from California.

Martin admits that one extremely wet season doesn't make a trend, and he has noticed other climate-connected changes that generally help his business. The spring last-frost date has shifted earlier over his twenty-nine years at Furmano, allowing him to move up planting dates without harming the transplants. In central Pennsylvania they used to begin planting limited acres of tomatoes cautiously on May 8. Now they plant many more acres earlier, without worrying about frost damage. "If this keeps going," quips Martin, "we'll be able to grow tomatoes in Maine."

A decade ago, the company joined Sysco's Sustainable/Integrated Pest Management Program, and Martin continues to track parameters such as energy, water, and pesticide use. To decrease diseases and improve the soil, Furmano grows tomatoes in a given field only once every four years, usually rotating with corn, soybeans, and snap beans. They also plant more cover crops—oats, rye, wheat, tillage radishes, and Austrian pea. With the warmer fall temperatures, Furmano Farms can sow these later than they used to, conserving and improving the soil in the process. Says Martin, "If the soybeans don't come off the field until October 20th, we're still getting in a cover crop. Our goal is to cover all the fields we grow vegetables on and try to cover everything else we can."

assessing perennial varieties that don't need annual replanting, and shifting planting times to avoid excessive temperatures. Entire regions must cooperate to keep these globally vital foods.

Now let's clear the table and welcome the grand finale—desserts and coffee.

Dessert and Coffee

Humans have evolved to prefer sweet flavors, and desserts satisfy that natural desire for sweet foods that provide energy and essential nutrients.[1] Fancy cakes and confections also stir our imaginations. Dessert is the highlight of the meal, a true celebration of chemistry and food that turns milk, sugar, chocolate, eggs, and other simple ingredients into dishes that delight the eyes as well as the taste buds.

"Confection" originally described a mixture to enhance health, and fourteenth-century curatives were sweetened to make them more palatable—in other words, people added sugar to "help the medicine go down." In the sixteenth century, hosts gave their dinner guests candied fruits and nuts after a large banquet. Modern sweets owe much to sugarcane, but ancient Egyptians cultivated honey, and their neighbors satisfied sweet cravings with dried figs, dates, and other fruits.[2]

Warming winters, surprise frosts, and water shortages are affecting nuts and fruits, and weather extremes, changes in precipitation patterns, and higher temperatures have an effect on dessert flavorings, among them banana, coconut, nutmeg, and saffron.

In this chapter we explore how climate change affects a few key dessert ingredients—sugar, chocolate, dairy, and vanilla—and we top off the meal with an item many of us consider essential, coffee.

Sugar

Your fair discourse hath been as sugar,
Making the hard way sweet and délectable.

WILLIAM SHAKESPEARE, *RICHARD II*

At the heart of dessert lies sugar, typically derived from sugarcane or sugar beets. Produced since the fifth century BC in northern India and the third century BC in China, the crystalline white grains spread to Persia and then Europe over several centuries. Later, explorers brought sugarcane to the Americas, where it flourished, becoming the New World's first profitable agricultural export. This expansion of the sugar industry, however, came at an extraordinary human cost. An untold number of indigenous lives were destroyed and eleven million Africans enslaved, all to fuel the "age of sugar."[3]

The unique chemical properties of sugar make it perfect for desserts. Heated to various temperatures, it helps create the unique textures of

Sugarcane

Different forms of sugar are created as it is heated, especially important when making candy.

chewy fudge (when mixed with chocolate and condensed milk), marzipan (when combined with almond paste), or long-lasting, glassy peppermint sticks.

Sugar is removed from cane or beets in a series of solid-to-liquid separations. Once refined, beet sugar is indistinguishable from cane sugar, and the two have nearly identical chemical makeups. Sugarcane thrives in tropical climates, whereas sugar beets grow best in temperate regions.

Where Does Sugar Come From?

Globally, growers produce nearly 2 billion tons (1.9 billion metric tons) of sugarcane annually.[4] Brazil's production is the highest in the world, with India second and China, Thailand, Pakistan, and Mexico also growing large quantities.[5] Florida, Texas, Louisiana, and Hawaii grow most of the US sugarcane crop, valued at $960 million annually.[6] Russia produces the most sugar beets, over 57 million tons (52 million metric tons) in 2017.[7]

The red beet (left), which we often eat, is dwarfed by a sugar beet (right).

What Is Changing for Sugar?

We don't know as much as we need to about how climate change will affect sugarcane, but we know impacts will vary greatly by region. For example, less rainfall will reduce sugarcane production in Brazil's northeast while higher temperatures in southeastern Brazil will likely increase yields there. Research does show that sugarcane is extremely sensitive to drought and tropical cyclones. When a 2006 cyclone hit northeastern Queensland, Australia, estimated crop losses were about 250,000 tn. (227,000 t).[8] In 2012 and 2013, orange rust thrived during warmer winters and higher-than-average humidity in Florida, hurting local sugarcane crops and increasing the costs of control.[9] In Europe, warmer springs will likely increase sugar beet yields in some areas, while drought-caused losses could double where drought is already a problem and extend into new areas. Losses from drought in western and central Europe could reach 18% by 2050, if not sooner.[10]

It's not just sugar that's sweet

About 6% of the US corn crop goes to make high-fructose corn syrup. The United States produces by far the most in the world—8.3 million tons (7.5 million metric tons) in 2017[a]—and uses the corn starch-based sweetener in beverages, processed foods, bakery goods, confections, and dairy products.

Other options, often referred to as "natural sweeteners," include maple syrup, honey, stevia, molasses (from sugarcane), agave nectar, date nectar, brown rice syrup, and sorghum syrup. The United States in recent years has produced about 4.2 million gallons (16 million liters) of maple syrup[b] and 149 million pounds (68 million kilograms) of honey.[c] The US market volume of stevia, a plant-derived sweetener two hundred times sweeter than refined sugar, was expected to be valued at $60 million in 2020, growing to $84 million by 2024.[d] Globally, natural sweeteners are a very large business that could be worth $37 billion by 2025.[e]

Climate change will affect plant-based sweeteners. For example, Canada produces most of the world's supply of maple syrup, but syrup makers in the northeastern and midwestern United States also collect sap in late winter and early spring. Warmer March temperatures have pushed tree-tapping earlier, and warmer spring and summer temperatures lower sugar content the following tapping season.[f] Scientists predict that future sap collection will be about one month earlier than it is currently and that the best sap flow will be about 250 mi. (400 km) farther north by 2100.[g]

The honeybee, the main source of honey, already faces challenges from pesticides and new diseases. Climate change will further exacerbate

(Continued on next page)

Chocolate

Food of the gods

TRANSLATION OF *THEOBROMA*

Like wine and coffee, chocolate comes in very distinct types, including shade grown, single varietal, domain specific, and fair trade. In 2015, people all over the world enjoyed about 8 million tons (7.3 million metric tons) of chocolate, some relishing the nostalgic comfort of a Hershey's

(Continued from previous page)

these stressors and add more for this critically important species.[h] Finally, stevia, grown mostly in China, Kenya, Paraguay, and the United States, will likely face increasing risks from climate change, but the research on specific impacts appears limited.[i]

---

[a] "Sugar and Sweeteners: Background," USDA Economic Research Service, updated August 20, 2019, https://www.ers.usda.gov/topics/crops/sugar-sweeteners/background/.

[b] "Total Maple Syrup Production in the United States from 2012 to 2019," Statista, released June 2019, https://www.statista.com/statistics/372073/us-maple-syrup-production-by-state/.

[c] "Production Volume of Honey in the United States from 2010 to 2018," Statista, released May 2019, https://www.statista.com/statistics/593656/us-honey-production-volume/.

[d] "Forecast Market Value of Stevia in the United States from 2013 to 2024," Statista, released October 2018, https://www.statista.com/statistics/937546/stevia-market-value-us/.

[e] "Value of the Natural Sweeteners Market Worldwide from 2017 to 2026," Statista, released January 2018, https://www.statista.com/statistics/958229/global-natural-sweeteners-market-value/.

[f] Kristina Stinson et al., "Climate Effects on the Culture and Ecology of Sugar Maple," Northeast Climate Adaptation Center, accessed March 24, 2020, https://necsc.umass.edu/projects/climate-effects-culture-and-ecology-sugar-maple.

[g] Joshua M. Rapp et al., "Finding the Sweet Spot: Shifting Optimal Climate for Maple Syrup Production in North America," abstract, *Forest Ecology and Management* 448 (September 2019):187, https://doi.org/10.1016/j.foreco.2019.05.045.

[h] Yves Le Conte and Maria Navajas, "Climate Change: Impact on Honey Bee Populations and Diseases," *Revue Scientifique et Technique (International Office of Epizootics)* 27 (September 1, 2008):499–510, https://apinz.org.nz/wp-content/uploads/2018/06/Le_Conte-et-NAvajas-OIE-ANG-copy-Srept-2008.pdf.

[i] "Where Does Stevia Come From?," *PureCircle Stevia Institute* (blog), accessed October 19, 2019, https://www.purecirclesteviainstitute.com/resources/infographics/stevia-facts/where-does-stevia-come-from/.

bar softened by a toasted marshmallow, others the bitter snap of a color-fully labeled 85% cocoa bar.[11]

Where Does Chocolate Come From?

Milk chocolate or dark, candy coated or peanut laden, this confection is made from the seeds of the cacao tree (*Theobroma cacao*). Botanists believe cacao originated in the Amazon or near Chiapas, Mexico.[12] Between 1300 and 1521, Maya and Aztec elites savored chocolate during ceremo-

Brazil's sugarcane expansion is sickly sweet

More than ninety countries grow sugarcane, the planet's number one crop by weight. The plant offers not only sugar but also bagasse fiber, which can be converted to ethanol to meet energy needs. In 2014, roughly 20% of global ethanol came from sugarcane, a number expected to increase to 26% in 2024. Sugarcane can also be used in plant-based plastics, paper, animal feed, and fertilizers.

Brazil, the world's top sugarcane producer, is expanding its plantations and encouraging further ethanol production from sugarcane, prompting concerns about deforestation, threats to ecological systems, and food security. Increases in production and rising temperatures are inadvertently helping rodents that breed and feed amidst the cane to spread hantavirus among themselves and to people.

These rodents prefer eating canes over the native vegetation. Warm temperatures encourage more rodent generations, raise their survival rates, and increase the time hantavirus remains viable. When workers and others living near the plantations inhale infected rodent excrement and saliva, they can develop hantavirus cardiopulmonary syndrome (HCPS). Fifty percent of patients with HCPS will die, and no vaccine is available.

Ecologist Paula Ribeiro Prist at the University of São Paulo and her colleagues at Columbia University in New York say burgeoning sugarcane plantations could put one-fifth more people at risk for developing HCPS, and a warmer climate could bump that to one-third more.[a]

---

[a] P. R. Prist et al., "Climate Change and Sugarcane Expansion Increase Hantavirus Infection Risk," abstract, *PLOS Neglected Tropical Diseases* 11, no. 7 (July 30, 2017):e0005705, https://doi.org/10.1371/journal.pntd.0005705.

nies such as weddings, and eventually conquistadors brought cacao beans to Spain. Over time, the luxurious treat spread widely.

Like coffee, cacao grows in a thin band near the equator. Roughly two million farmers in the small West African cocoa belt—from Sierra Leone to southern Cameroon—grow about 70% of the world's cacao and produce the highest volume of high-quality bulk cocoa, as opposed to specialty cocoa.[13]

Chocolate terminology

Cacao, cocoa, or chocolate? The terms are often confused. Cacao refers to the plant and the dried beans. Further bean processing—roasting, fat removal, and grinding—results in cocoa powder or cocoa solids. Chocolate is the final food product and often includes such ingredients as sugar, milk, and vanilla.

Globally, cacao production depends on five to six million farmers, most of whom farm fewer than 5 ac. (2 ha). Individual farmworkers earn under $2.00 per day, well below a living wage.[14] Cacao cultivation is very labor-intensive work and demands continuous attention. Twice a year over one hundred thousand small flowers bloom on each cacao tree and develop into small green pods. Using machetes and poles, growers hand-harvest mature pods, which are the size of a football and contain around thirty to forty-five beans. With proper care, cacao trees can usually produce pods for thirty years.

Most cacao is grown under direct sunlight in monocultures that, under perfect conditions, yield optimally. However, this cropping system often depletes the soil nutrients, which are not replaced.[15]

How Do Beans Become Chocolate?

Cacao beans are fermented to jump-start the flavors and aromas. Traditional small landholders ferment beans in a pile covered by banana leaves, whereas larger operations ferment up to 2 tn. (1.8 t) at once. Next, beans are dried, packed into burlap sacks, and sent to processors, who clean and roast them to darken the color and develop the flavor. Processors remove the shells to reveal roasted cocoa nibs, which they grind into a paste called chocolate liquor or cocoa mass. They ship the liquor to manufacturers who blend it with other ingredients such as sugar, vanilla, and milk. To make the final product smooth, chocolate makers mix and aerate the ingredients at high temperatures in a process called conching. Next, they temper and mold the chocolate, which is now ready for packing, shipping, and eating.

Cacao pods.

How Big Is the Chocolate Business?

Chocolate is the confectionary industry's most consumed and profitable product. In 2017, the global chocolate confectionary market was valued at $129 billion, and it's projected to climb over $187 billion by 2026.[16] In the same year, global cacao bean production reached 4.7 million tons (4.3 million metric tons).[17] Europeans enjoy more chocolate than do people in the United States, and the Swiss top the list, with the average person consuming nearly 19.5 lb. (8.8 kg) per year.[18]

How Is Cacao Production Changing?

The narrow equatorial cacao belt is likely to see higher temperatures and get drier, not a good combination for cacao plants.[19] West Africa has a long dry season that is becoming even drier, with annual rainfall there predicted to decrease up to 30% over the second half of the twenty-first century.[20] Although the region's cacao trees are well adapted to increasing temperature, water shortages threaten local production.

In other cacao-growing regions, increased droughts, pests, and diseases could hurt production. For example, a severe drought in Bahia, Brazil, during the 2015–2016 growing season led to 15% of the cacao trees dying and up to an 89% decrease in yields.[21] A fungus, frosty pod rot, is rapidly becoming the most destructive disease of cacao in South America. Changes in the global temperature could help it spread to other, previously unaffected regions.[22]

What Is Being Done to Save Our Chocolate?

Growers can help ensure a successful cacao crop by moving to higher and cooler elevations, but doing so may encroach on existing forest preserves.[23] Likewise, they can protect cacao trees by planting bananas, plantains, and other crops that provide overhead shade, reducing cacao leaf temperatures by up to 7°F (4°C).[24] Adding shade crops can give cacao farmers additional cash crops and timber plus conserve their soil. With another method, cabruca, farmers remove a few large trees from the rainforest to make space for cacao, avoiding disease and excessive heat while preserving the forest and storing carbon in the large rainforest trees.[25]

The cacao industry's survival also depends on making the cacao plant itself more resilient. Far from the nearest cacao operation, in the rainy, suburban town of Reading, England, the International Cocoa Quarantine Centre houses over four hundred varieties. Here scientists are assessing cacao's genetic diversity and developing disease-resistant varieties.[26] Other scientists now have the first proof of concept for using gene editing to increase pathogen resistance.[27]

In Turrialba, Costa Rica, scientists at the Tropical Agricultural Research and Higher Education Center are also working to create a cacao plant with increased disease resistance and yield and the same ability to produce high-quality, good-tasting chocolate. The center has collected twelve hundred clones, or identical twins, of cacao plants from tropical America and elsewhere. They partner with the International Cocoa Quarantine Centre to further enhance the diversity of the collection.[28]

The chocolate industry has begun to respond to climate change, stakeholder pressure, business needs, and larger corporate strategy shifts by establishing sustainability standards along the supply chain. In 2000, chocolate industry members created the World Cocoa Foundation, which focuses on sustainability, including farmer training and increasing global investment in crop production. The foundation helps smallholders enhance productivity, manage differently, and diversify to increase their resilience to climate change.[29] Since the early 2000s, Cadbury, Mars, Ferrero, and other companies have embraced third-party fair trade and sustainability certification. Lindt developed its own internal sustainability standard, and Nestlé uses third-party certification as part of a broader strategy.[30] In 2017, Mars announced an ambitious greenhouse gas emission reduction goal of 67% along its full value chain by 2050.[31] It has committed $1 billion to the effort.[32]

Dairy

Remember the bells of the Good Humor truck advertising ice cream on sweltering summer afternoons? People in the United States slurp a lot of ice cream annually. It's an $11 billion business supporting twenty-six thousand

jobs. Our favorites include vanilla, the most popular ice cream flavor, followed by chocolate and cookies 'n' cream.[33]

The United States is the world's leader in the production of cow's milk.[34] More than nine million cows produce enough milk annually to fill about thirty-nine thousand Olympic-sized swimming pools, putting dairy in the country's five leading agricultural commodities. Although every US state produces milk, California, Wisconsin, New York, Idaho and Texas top the list.[35]

Dairy farms have changed dramatically since 1970, when the United States had 648,000 dairy farms, averaging nineteen cows per farm.[36] When milk revenues didn't keep pace with farming costs, dairy farmers either relied on off-farm income, became larger and more efficient, or sold out. Currently about 40,000 farms remain, and while many are big—in 2018 average herd size was over 1,000 in nine states—farms in leading dairy states such as New York and Wisconsin averaged about 150 cows each.[37]

As the number of dairy farms shrank, the milk produced per cow increased owing to improvements in breeding, feeding, technology, facilities, and overall herd management.

How Is the Dairy Market Changing?

Of more than 140 million milk cows worldwide, some are considered holy, many pull plows, and most provide humans with sustenance, converting plants into one of the most nutritious foods that's rich in protein, carbohydrates, vitamins, and minerals such as calcium, magnesium, and zinc.[38]

In developing countries, dairy fulfills only 4% of caloric intake, in developed countries, 14%.[39] By 2050, projected global use of dairy will likely increase substantially because of an overall rise in the standard of living.[40] China will drive some of that future demand. In 1961, the Chinese drank just over 4 lb. (2 kg) of milk per capita, but demand could jump to 180 lb. (82 kg) a year by 2050. Currently China is the globe's top importer of milk, and it's gearing up domestic production, which could increase nitrogen pollution, greenhouse gases, and land and water use. Adoption of more efficient milk and feed production tactics, however, could mitigate these increases.[41]

Drinking cow's milk gained popularity in the United States in the first decades of the twentieth century, coinciding with mass pasteurization and the temperance movement's push to replace alcohol with a healthy beverage.[42] However, consumption has been declining for several decades, in part because of competition from other beverages such as carbonated soft drinks, bottled water, and milk alternatives.[43] In addition to producing fresh milk, the dairy industry churns out butter, dry milk powders, evaporated and condensed milk, yogurt, whey products, hundreds of soft, semihard, and hard cheeses—and, of course, ice cream.

How Does Dairy Contribute to Climate Change?

The total dairy sector contributes about 2% of all greenhouse gases emitted in the United States. This estimate includes the entire food chain, from planting feed corn and storing manure to processing the milk, refrigerating the products, eating—and wasting—dairy foods at home, and toting away the empty milk carton. Overall, the largest contributors of greenhouse gasses from the dairy sector include cows, which produce methane (like beef animals do), manure management, and feed production.[44] Worldwide, more efficient production and other practices are lowering the average

Cheeses and desserts

Cheese can be found throughout the meal as a delightful garnish, the centerpiece, or even a delectable dessert. The global fascination and admiration for cheese stems from the diversity and artisanship associated with making it. There are thousands of different types of cheese.

Cheese does not need to be eaten as is, however. Many cheesecake devotees enjoy the classic dessert either baked or unbaked. And what would a cannoli be without ricotta cheese, or red velvet cake without cream cheese frosting? Baked brie with fresh berries, walnuts, and honey can satisfy a sweet tooth, or try a fruit tart with a smidge of goat cheese. Cheese can be folded into pudding, mousse, and panna cotta, as well as cakes and frostings. The possibilities are nearly endless: a simple ricotta with a drizzle of honey is a sweet and simple dessert. Or perhaps a vibrant blue cheese from Spain with a hint of spice, enhanced with quince paste and Marcona almonds. The combination of sweet and savory is a lovely end to a meal, and stacks of cheese rounds are even replacing traditional wedding cakes as a more interesting celebration piece.

greenhouse gas emissions per standardized unit of milk. Emissions in North America are about half of the global average.[45]

When US milk production exceeds consumption, milk is either converted into dry milk powder, butter, or cheese to extend its shelf life, or it's disposed of. During the first eight months of 2016, more than 40 million gallons (151 million liters) of excess milk were poured onto fields and into manure storages or discarded at processing plants. That was about 40% more than the amount of milk disposed of during each of the preceding fifteen years.[46] Dairy farmers bear the brunt of falling milk prices and can't quickly "turn off the cows." Plenty simply go out of business, as seven hundred did in Wisconsin in 2018.[47]

How Is the Dairy Industry Changing?

In addition to predicted warming, many US farmers could also struggle with water shortages for crop production and livestock.[48] A lactating dairy

Milk alternatives sprouting up everywhere

You've seen the cartons of soy, almond, cashew, and coconut beverages squeezing their way into the dairy case. Half of US shoppers will put one of these in their shopping cart this year.[a] Globally, industry revenues from alternative milks are projected to exceed $38 billion by 2024.[b] Soy commands the largest market value worldwide, followed by almond and coconut.[c]

These nondairy alternatives might seem like a fad, but some of them have been around for centuries. Almond milk is mentioned in a Middle Eastern cookbook from 1226, and the Chinese probably began making soy milk in the 1600s; it became known in the United States in the early 1900s.[d]

Why the popularity now? About one in thirteen US children under five is allergic to milk, and thirty to fifty million US adults have trouble digesting the lactose in milk.[e] A quarter of consumers buy plant-based beverages to avoid dairy allergies. Roughly two-thirds think they taste better and are healthier than cow's milk, and others believe their production is more humane than dairy products.[f]

Only soy milk contains as much protein as cow's milk, but how readily it is absorbed by the body is still being studied. Almond and rice milk contain about 13% as much protein as soy milk.[g] Although most plant milks are fortified with calcium and other minerals, soy alone is considered a nutritionally acceptable substitute for cow's milk, especially for children.[h]

(Continued on next page)

cow drinks up to 50 gal. (190 L) of water daily and much more when she is heat stressed.[49]

Modern US dairy barns are normally well ventilated and unheated, even during winter, because cows like it cool and dry and generate a lot of heat. Their ideal temperature range is from just below freezing to about 65°F (18°C). Under high humidity, even with moderate temperatures, their metabolism increases to combat heat stress.[50] They drink and respire more but eat less, which lowers milk production. Some heat-stressed cows produce milk with less protein, lactose, and fat. Cows' immune systems can weaken in hot conditions—possibly necessitating more medications—and

---

(Continued from previous page)

In relation to climate change, a 2018 study found that production of cow's milk generated about three times more greenhouse gases than rice, soy, oat, or almond milk. In addition, it required several times the amount of land and used more water.[i]

---

[a] Oliver Franklin-Wallis, "White Gold: The Unstoppable Rise of Alternative Milks," *Guardian*, January 29, 2019, https://www.theguardian.com/news/2019/jan/29/white-gold -the-unstoppable-rise-of-alternative-milks-oat-soy-rice-coconut-plant.

[b] "The Global Non-Dairy Milk Market Is Projected to Reach Revenues of More Than $38 Billion by 2024," MarketWatch, March 26, 2019, https://www.market watch.com/press-release/the-global-non-dairy-milk-market-is-projected-to-reach -revenues-of-more-than-38-billion-by-2024-2019-03-26.

[c] "Market Value of Dairy Milk Alternatives Worldwide in 2019, by Category," Statista, last modified October 23, 2019, https://www.statista.com/statistics/693015/dairy-alter natives-global-sales-value-by-category/.

[d] William Shurtleff and Akiko Aoyagi, *History of Soymilk and Other Non-dairy Milks (1226 to 2013)* (Lafayette, CA: Soy Info Center, 2013), http://www.soyinfocenter.com/books/166.

[e] "Facts and Statistics: Food Allergy Research & Education," Food Allergy Research and Education, accessed March 15, 2019, https://www.foodallergy.org/resources/facts -and-statistics; "Lactose Intolerance," Genetics Home Reference, National Institutes of Health National Library of Medicine, March 12, 2019, https://ghr.nlm.nih.gov/condition/lac tose-intolerance.

[f] "Reasons for Buying Plant-Based Milks among Consumers in the United States as of Q4 2015," Statista, released January 2016, https://www.statista.com/statistics/516308 /plant-based-milks-reasons-to-buy-us-consumers/.

[g] Meagan Bridges, "Moo-Ove Over, Cow's Milk: The Rise of Plant-Based Dairy Alter-natives," *Practical Gastroenterology* 7 (2018):153–59, https://med.virginia.edu/ginutri tion/wp-content/uploads/sites/199/2014/06/January-18-Milk-Alternatives.pdf.

[h] Margaret Schuster et al., "Comparison of the Nutrient Content of Cow's Milk and Non-dairy Milk Alternatives," abstract, *Nutrition Today* 53, no. 4 (August 7, 2018):153, https:// doi.org/10.1097/NT.0000000000000284.

[i] J. Poore and T. Nemecek, "Reducing Food's Environmental Impacts through Producers and Consumers," *Science* 360, no. 6392 (June 1, 2018):988, https://doi.org/10.1126/science .aaq0216; Clara Guibourg and Helen Briggs, "Climate Change: Which Vegan Milks Are Best?," BBC News, February 22, 2019, https://www.bbc.com/news/science-environment-46654042.

---

their fertility rates drop; calves can be smaller and less vigorous.[51] Heat stress effects can last weeks to months.

Location is already affecting financial success. Dairies in warmer regions generally spend more money keeping their cows cool than those in cooler regions. Changing housing and altering practices to prevent climate-induced heat stress in the future might help, but the US industry could lose $106 to $269 million annually.[52]

A warmer climate might not mean total gloom for dairies, however. Longer growing seasons could allow farmers to plant more productive feed varieties and grow two or more crops on the same land, thus reaping more per unit of land than ever before.[53]

How Can Cows Beat the Heat?

Most cows in the United States are kept cool with misting and sprinklers. Farmers also reduce overcrowding in barns, minimize time in confined holding areas, use fans and natural circulation, and make sure that pastured cows have access to trees or shade structures. Some farmers use a combination of sand, buried plastic radiators, and water-cooled pads as bedding material to help alleviate heat stress. More expensive measures include redesigning buildings and installing high-pressure fogging systems.[54]

Cows experiencing heat stress use feed less efficiently because digestion and nutrient absorption bog down. To beat the heat, farmers can "feed them like queens," providing fresh, high-quality forages at cooler times of the day and adjusting feed mixes to counteract the cows' losses.[55]

Another approach to help keep the milk flowing is breeding productive, more heat-tolerant animals that, in an ideal scenario, would produce less methane. For such breeding, researchers start with breeds that already produce well, which are mostly breeds from temperate climates.[56] Senepol cattle have characteristics that help them deal with tropical temperatures: a sleek coat with shorter hairs, fewer hair follicles, and larger sweat glands than other breeds.[57] Recent studies at the University of Florida with mixed Holstein–Senepol "slick" cows showed they could regulate their heat better and produce more milk in the summer than nonslick cows. When considering traits for their dairy cows, producers in the future might be able to select for resistance to heat stress.[58]

How Can Farmers Reduce Greenhouse Gases and Cope with a Warming Climate?

To curb greenhouse gases, farmers have several options. They can seal manure storage tanks so the gases don't escape and then combust those gases to produce energy. As with beef cattle, farmers can manage pastures well, practice silvopasture to help maintain soil carbon, and improve forages. In the future they may be able to select breeds with gut microbes that produce less methane.[59]

How efficiently dairy cows convert feed to milk directly affects their contributions to greenhouse gases. Farmers can lower enteric methane emissions by modifying the animals' diets, but they must do so carefully. Supplementing a cow's diet with fats and oils can reduce enteric methane and is promising as long as digestibility is not affected.[60] In the near future farmers might be feeding cows 3-nitrooxypropanol, a product made by a Dutch company that reduces methane production more than 20% without compromising milk productivity or quality.[61] Other approaches may be to feed dairy cattle seaweed or garlic. A recent study showed that including as little as 1% seaweed in the diet could reduce methane production by 50%.[62] Cows that were fed garlic produced half as much methane.[63]

Larger dairies can take advantage of anaerobic digesters. Anaerobic digesters can significantly reduce greenhouse gas emissions from dairies, adapting from municipal wastewater treatment the concept of creating energy while digesting solids.[64] Liquid manure is confined to a sealed tank where anaerobic ("without oxygen") microbes produce biogas, which is about half methane and half carbon dioxide.

Anaerobic digesters help farmers get additional value from manure and reduce the cost of transporting it. Biogas from anaerobic digesters can be used on site, piped to an internal combustion engine and generator to produce electricity, or sold to a biomethane processing facility.[65] Post-digested, separated solids can be used as fertilizer or cow bedding. In addition, anaerobic digestion partially reduces the volume of manure solids, helps keep pollutants out of waterways, kills pathogens, reduces odor, and helps farmers comply with environmental regulations for manure handling.[66] In 2018, anaerobic digester systems on US livestock farms (dairy, cattle, swine, and poultry combined) generated the equivalent of about 1.1 million megawatt-hours of electricity.[67] (See illustration on the following page.)

What's not to like? First, they're expensive. Some cost-sharing or reimbursement programs are available through the US Department of Agriculture, the Rural Energy for America Program, and federal funding channeled through states. But federal support is still in its infancy, and systems can cost a million dollars or more to install.[68] Co-digesting with municipal food wastes can generate tipping fees and help the bottom line,[69] but it's still hard for anaerobic digesters to compete with historically low energy prices in the

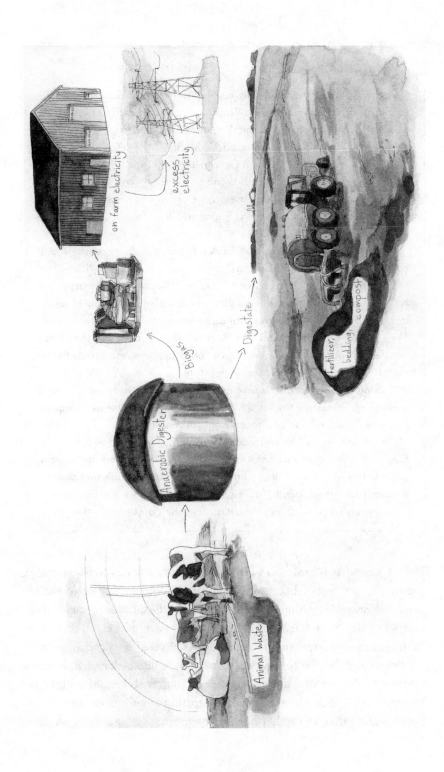

Ben & Jerry's serves up incentives

An activist organization that sells ice cream? That's how Ben & Jerry's describes itself, according to Dave Rapaport, long an activist himself. He now heads the global social mission team for Ben & Jerry's, a Unilever subsidiary based in South Burlington, Vermont, with retail outlets in about forty countries. His team leads a climate justice campaign and promotes sustainability, human rights and dignity, and environmental protection. "We drive those values through the way we operate as a business and also the way we use our influence," says Rapaport. "We really believe that business can be a force for good."

Dairy accounts for about 50% of Ben & Jerry's carbon emissions, and so the company developed Caring Dairy to spur sustainability. "It's geared toward incremental improvement," explains Rapaport. Farmers plant cover crops, monitor the nutrients they add to their no-till or low-till fields, track greenhouse gases, and manage manure to minimize emissions. They receive a premium from Ben & Jerry's that increases as they make progress. Rapaport says Caring Dairy farms are now responsible for almost half of cover-cropped acres in the state.

Caring Dairy's partner agencies also work intensively with farmers through meetings and workshops. Rapaport's goal is for farms to sequester enough carbon that they become net carbon sinks.

"It is challenging," he concedes. "With things so tough with farmers, they're not always able to make the investments." Separators that remove solids before manure is dumped into storage pits and aerobic and anaerobic digesters would speed progress, and Ben & Jerry's is partnering with Vermont-based NativeEnergy to cut greenhouse gases even further.

United States. As of 2019, 248 farms in the United States were generating energy with anaerobic digesters.[70] California expects to have up to 120 anaerobic digesters operating by 2022.[71] The state's East Bay Municipal Utility District now owns eleven anaerobic digesters as part of a system that draws from households, farms, and food processors, exceeding its energy needs.[72]

If cows and the plants they eat can't handle climate-change heat, more farmers might consider milk goats, but only after some major kinks get worked out. Worldwide, the carbon footprint for milk from small ruminants is more than twice that of milk from cows.[73] Not only do sheep—

If the United States gets really hot, there's always camel milk

Boasting high amounts of calcium, iron, protein, and some vitamins, camel milk has entered the US market. It has less cholesterol than cow's milk and fewer proteins that contain common milk allergens.[a] In 2014, the United States had about five thousand camels, and a portion were milked for commercial sales. Get out your wallet, though—the milk is expensive. A pint (0.5 L) of camel milk retailed for about $18.00 in 2018.[b]

---

[a] Kim Ann Zimmermann, "Camel Milk: Nutrition Facts, Risks & Benefits," Live Science, February 3, 2016, https://www.livescience.com/53579-camel-milk.html.

[b] "Camel Milk," Wikipedia, accessed July 12, 2019, https://en.wikipedia.org/wiki/Camel_milk; Sarah Lazarus, "Saudi Entrepreneur and Amish Farmers Bring Camel Milk to US," CNN, updated November 25, 2018, https://www.cnn.com/2018/11/25/health/camel-milk-in-the-us-intl/index.html.

and to a lesser degree, goats—produce more methane per standardized unit of body weight than cows, but climate-friendly protocols for managing them are still poorly developed.[74] That's unfortunate, because goats in particular tolerate harsher conditions than do dairy cows. They are adept grazers and digest food low in protein.[75]

Globally, sheep and goats together contribute about 5% to the milk supply.[76] Sheep milk, which is high in riboflavin, thiamine, niacin, and other vitamins, is used primarily for making cheeses like Roquefort, feta, and pecorino. It is very important in Middle Eastern and Mediterranean countries.[77]

Goats feature prominently in tropical Asian and African agriculture.[78] From 1981 to 2001 the number of dairy goats increased over 50% in developing countries and a surprising 17% in developed countries.[79]

The 373,000 dairy goats in the United States represent a slowly growing industry.[80] One doe, depending on the breed, can produce up to 5 qt. (4.7 L) of milk per day. Compared with cow's milk, goat milk has equivalent amounts of essential amino acids and more calcium, magnesium, potassium, phosphorus, niacin, and thiamine. People with allergies to cow's milk can often drink goat milk without problems.[81]

Will sheep and goats ever displace the highly productive dairy cow? Unlikely, but some producers might consider diversifying if market demand

and climate changes merit it. Researchers and economists might someday recommend this pathway, among others, to increase the sustainability of the US dairy industry.

Vanilla

While "plain vanilla" sounds dull, vanilla is an exotic and highly valuable spice with over two hundred compounds that contribute to its aroma and flavor.[82] The vanilla beans most people are familiar with are the fruits of a climbing orchid, *Vanilla planifolia*, a species native to the Caribbean and South and Central America.

Vanilla's origin story is similar to chocolate's, but Mayan elites may have prized vanilla beans more. The beans were added to the chocolate drink shared with the Spanish conquistador Cortés, and soon after, vanilla made its way to Europe to be used in perfumes and desserts. Despite its value as a spice, production was essentially impossible outside its native habitat because pollination depended on *Melipona* bees. A twelve-year-old former slave, Edmond Albius, overcame this obstacle on Réunion Island in the Indian Ocean.[83] His ingenious method of hand pollinating vanilla blossoms paved the way for the vanilla industry and boosted production of one of today's most expensive spices, second only to saffron. Today, Réunion Island produces some of the world's best vanilla.

How Is Vanilla Produced?

Vanilla production is labor intensive, which adds to its high cost. Vanilla prefers temperatures between 70°F (21°C) and 90°F (32°C), lots of rain, and high humidity; it thrives in Madagascar and Indonesia. The plants grow about three years before they're ready for pollination and producing fruit. Workers pollinate by hand, using a pointed bamboo splinter to open the flower and transfer pollen from the anther to the stigma. They have to work quickly because plants bloom for only two months and individual flowers bloom for only a day. Fruits are ready for harvesting six to nine months later and roughly six hundred hand-pollinated blossoms will yield 2.2 lb. (1 kg) of beans.

The most widely used process for curing vanilla beans starts with soaking them briefly in hot water, placing them in insulated crates to sweat for several hours, then drying them slowly for weeks and storing them in boxes for conditioning. The cured beans may then be soaked in alcohol for twelve days to extract the flavor.

Globally, the export value of vanilla is about $1.3 billion, with the United States importing approximately 43% of the world's natural vanilla.[84] Demand will likely increase for natural vanilla, given consumer preference for more natural foods. Large companies such as Nestlé and Hershey's have committed to using natural ingredients in their products.[85]

Despite the large natural vanilla industry, most people are familiar with synthesized vanilla, which provides 99% of "vanilla" flavoring.[86] Synthetic vanilla, at $7 per pound ($15 per kilogram), is much cheaper than natural vanilla, which in 2018 was as high as $227 per pound ($500 per kilogram) for first-grade extraction.[87] Both natural and synthetic vanilla find their way into many products but are primarily used to flavor chocolate, beverages, and, of course, desserts. They are also found in soaps, perfumes, cosmetics, pharmaceuticals, and candles.

What's Changing for Vanilla?

As demand for natural vanilla increases, a changing climate poses challenges. Like cacao, vanilla is grown mainly by smallholder farmers who are extremely vulnerable to system shocks, such as extreme weather, pests, diseases, and increasing prices for agricultural inputs.[88]

Madagascar, the source of most natural vanilla, has experienced increasing temperatures, heat waves, drought, floods, and cyclones. In 2017, Cyclone Enawo, the strongest to hit Madagascar in thirteen years, devastated Antalaha and Sambava, the country's largest vanilla-producing regions. It destroyed 90% to 100% of the vanilla crops in Antalaha and 80% in Sambava, together almost 30% of the annual global supply, driving up prices by nearly 350%.[89]

What Is Being Done for Vanilla?

Centuries of cultivation have shown that vanilla is unlikely to be grown on a large scale in new locations; therefore high-producing countries such as Indonesia and Madagascar need adaptive measures to maintain vanilla production. Several companies, including Danone, Firmenich, and Mars, have partnered with Livelihoods Fund for Family Farming to teach vanilla farmers more sustainable production methods.[90] One is agroforestry, where trees and shrubs are grown with crops, which has been shown to help maintain soil moisture, prevent erosion, and minimize temperature extremes.[91]

Although chemically synthesized vanilla dominates the market, researchers have been turning to bioengineered microorganisms to produce compounds that naturally occur in vanilla beans, producing a flavor closer than the synthetic version to natural vanilla.[92] United States legislation has permitted this flavor to be labeled "natural."[93] This high-tech version is a promising alternative, should the conditions for growing vanilla become too unfavorable or costly.

Coffee

On the eighth day, God created coffee.

ANONYMOUS

Our imaginary meal ends with coffee, a beverage loved worldwide. It not only tastes wonderful, it wakes us up and keeps us healthy. Moderate coffee consumption reduces the risk of Parkinson's disease, Alzheimer's, and memory loss due to aging and improves our mood, to name just a few benefits.[94]

Whether we call it joe, java, brain juice, morning mud, battery acid, or whatever, many of us can't imagine waking up without our coffee.

Legend has it that an Ethiopian goat herder noticed his flock became especially energetic after munching berries from a particular tree. The herder took some berries to the local monastery, where the abbot made them into a drink and found he could stay awake during the evening's long prayer session. From there, the story and beans spread. When coffee replaced tea in the colonies after the Boston Tea Party in 1773, Thomas Jefferson evidently described it as the favorite drink of the civilized world.

Where Does Coffee Come From?

All coffee is grown in the sunny equatorial belt, where the rich soils, temperatures, and annual rainfall have been just right. Global production in 2018 was 21 billion pounds (9.4 billion kilograms), with 40% from Brazil and 20% from Vietnam.[95] Other major producers include Colombia, Indonesia, and Ethiopia.

Arabica coffee, one of the two main types, constitutes about 70% of global java and is considered the best tasting—naturally mild, aromatic, and infused with varied flavors. Plants do best in temperatures between 64° and 70°F (18° and 21°C) and are typically grown at higher altitudes.

Robusta coffee, harsher, somewhat bitter, and with much more caffeine, is primarily grown for blends and instant coffee. It thrives in warmer conditions than Arabica. Neither type can handle frosts, and both require about 60 in. (150 cm) of rainfall per year. Altitude and the timing and length of the dry and rainy seasons affect the duration of flowering, number of harvests, and the method used to dry the beans. Coffee farmers generally put production and resistance to pests and diseases ahead of coffee quality when they choose which of several varieties to grow, though they also consider yield, cup quality, likelihood of fruit drop (important in areas of high wind and rain), and intensity of care and fertilization.

How Do Beans Become a Cup of Coffee?

Coffee growers plant coffee seeds (unprocessed beans), raise the seedlings, and then transplant them during the wet season. The tree needs three to four years before it can produce fruit, or "coffee cherries." An individual tree can live up to one hundred years but is most productive between the ages of seven and twenty, producing about 10 lb. (4.5 kg) of coffee cherries or 2 lb. (0.9 kg) of beans per year.[96] Farmers harvest the cherries once or twice a year, either by machine (on flatland) or by hand (on slopes). Hand picking is very hard work, and pickers are often paid by the pound. With plunging coffee prices in 2019, coffee producers in Vietnam and Brazil were mechanizing their operations and increasing productivity well above competing areas in Africa, Central America, and Colombia.[97]

Once picked, the coffee cherries must be dried quickly to minimize rot, so where water is limited farmers spread thin layers in large sunny areas. Over several weeks, they turn the cherries several times daily, covering them during rains and at night. A more water-intensive method removes the outer layers and places the cherries in fermentation containers for a few days to dissolve more of the coverings. The resulting beans are then dried and turned regularly under the sun or in heated tumblers.

Once they have removed all outer layers, workers polish, grade, and sort the beans by hand or machine. They bag this "green coffee" or place it in bulk containers for worldwide distribution. The roasting process typically occurs in the importing country and involves heating the beans to 400°F (204°C), which releases the oils and produces the characteristic flavors and aromas. Ground and brewed, the coffee is ready to enjoy either in our homes or in one of the millions of coffee shops and restaurants worldwide.

Who Cares?

Globally, twenty-five million coffee farmers supply the world with java. Over 120 million people in more than seventy countries depend on some aspect of the coffee value chain—growing, processing, trading, roasting, or serving it—for their livelihoods.[98]

The United States is the largest importer of coffee. Nationally, coffee provides more than 1.7 million jobs and is the most commonly consumed beverage. With over 60% of the population drinking it each day, the total economic impact is $225 billion, mostly in the food service industry.[99] Per capita consumption in the United States is 30 gal. (115 L) per year, but it's over twice that in the Netherlands.[100]

Drinking coffee can be a pleasurable, multisensory experience affected by the type of coffee, how it is prepared, what is added, and even the container

The many reasons why people in the United States drink coffee. ("Why Do You Usually Drink Coffee?" Statista, released February 2017, https://www.statista .com/statistics/320363/us-consumers—reasons-for-drinking-coffee/.)

in which it is served. It is an important social lubricant in our culture, and for centuries coffeehouses have been social hubs. Today, they're nearly everywhere. Starbucks, for example, had over thirty thousand stores around the globe in 2019.[101]

What Is Changing for Coffee?

Our changing climate is affecting almost every aspect of coffee production.[102] Coffee is sensitive to even a small increase in temperature and, like many crops, is more susceptible at certain stages of growth and development. Just a little warming at the wrong time can reduce yield, flavor, and aroma. In Tanzania, where about 2.5 million people depend on coffee for a livelihood, increases in nighttime temperatures since the 1960s have already caused yields to drop, and severe declines are expected as conditions continue to warm.[103] In parts of Mexico, increasing temperatures could reduce coffee production by over 30%, making it unviable in the 2020s.[104] With continued climate change, the world's coffee production area will likely be cut in half by 2050.[105] Scientists reported in 2019 that at least sixty wild coffee species, critical sources of traits for climate change resilience and pest resistance, might go extinct.[106] However, increasing levels

What's up with tea?

Tea is a $10 billion industry and the second-most popular drink in the world. Typed according to how it's processed—black, oolong, Masala, pu'erh, and green—tea is, like coffee, changing. Shifts in temperature, rainfall, and season length can affect not only yields but also the hundreds of unique chemicals in teas that determine their potential health benefits and flavors.[a] Because of these changes the livelihoods of over a million workers and many small landholders are at increasing risk in Assam, a state in northeast India that produces almost 20% of the world's tea.[b] Similar challenges face China, another major producer of tea. Tea farmers, however, are conserving soil, managing water better, and providing shade for plants to ensure that tea is here to stay.

---

[a] J. M. A. Duncan et al., "Observing Climate Impacts on Tea Yield in Assam, India," abstract, *Applied Geography* 77 (December 2016):64, https://doi.org/10.1016/j.apgeog.2016.10.004; Anna Nowogrodzki, "How Climate Change Might Affect Tea," *Nature* 566 (February 6, 2019):S10–11, https://doi.org/10.1038/d41586-019-00399-0.
[b] K. R. Dikshit and J. K. Dikshit, *North-East India: Land, People and Economy* (New York: Springer, 2013), 262.

of carbon dioxide in the atmosphere may mitigate some of the negative impacts of warming and drought. Some scientists suggest that the overall impacts may be less severe than assumed and that more research is needed to assess how coffee will respond to the current and anticipated changes.[107]

Globally, climate changes are also resulting in more disease and pest problems. Coffee leaf rust, a fungal disease, thrives under warmer conditions and is advancing into higher elevations as they heat up. Crop losses in 2012 and 2013 in Colombia and Central America ranged from 10% to 31% due to the rust, directly affecting thousands of harvesters and small landholders.[108] Most harmful is the coffee berry borer, a beetle that is also spreading into new areas as they become hotter, such as higher elevations and outward from the equator.[109]

In the coming decades, growers will likely find coffee harder to grow in existing production areas. Millions of small landholders across the globe could be out of business, along with millions of others along the coffee value chain. With a greatly diminished and more volatile supply, prices can only go up.

If coffee trees get roasted

In Brooktondale, New York, inside a quaint wooden building with a copper-horse weathervane, Jesse Harriott is cooling some of the 75,000 lb. (34,000 kg) of coffee beans he'll roast this year. He pours them into bags stamped Copper Horse Coffee and named for aromas from his childhood, such as Rumble Pony and Carriage House. As a very small proprietor in the coffee world, he keeps one eye on the supply of beans and the other on the global climate. For instance, a frost in Brazil would lead to fluctuations in both supply and price. Coffee plants are so sensitive to temperature, he says, that they can detect a change of 1°F (0.6°C).

Harriott buys only specialty-grade green Arabica beans, which he believes are delectable because of the long fruit maturation and resulting complex sugars. "In Central and South America, we've seen coffee leaf rust affecting 40% of the crop," he explains. Organic growers have limited options for controlling the disease. Some pull out infected plants, but since coffee is a perennial, says Harriott, doing so means "you reap nothing the next year."

Harriott knows too about the balance between the rainy and dry seasons and how rain causes mold or prevents growers from finishing coffee cherries on open patios.

Arabica must be grown at least 3,200 ft. (975 m) above sea level. He buys some of his coffee from higher-altitude farms and wonders whether he should be looking for higher and higher altitude coffees. Harriott also wonders whether coffee grown in the 6,500 ft. (2,000 m) range can continue to move up. So he doesn't put all of his beans in one basket; in a few countries he's chosen a 6,200 ft. high farm just to hedge his bets.

"I think climate change is in a lot of people's minds," says Harriott. But he's not apocalyptic. "When someone makes a statement that in fifty years coffee will be gone, that's questionable to me."

What Is Being Done to Save Our Coffee?

Coffee producers in developing countries are not well prepared for a rapidly warming world. Some major coffee companies now train farmers to adapt to new conditions and become more resilient.[110] Farmers are planting shade trees among coffee trees to buffer temperature extremes and enhance the habitat for birds that feed on the coffee berry borer.[111] Other options include choosing more pest- and heat-tolerant varieties and relocating upslope where feasible. Research in Ethiopia shows that relocation could increase coffee acreage there fourfold, but efforts would need to align with forest conservation practices.[112]

Growing conditions have changed so much and coffee rust has become such a problem in low-altitude areas of Central America that coffee production is no longer an option, and some farmers are switching to cacao, which thrives in the warmer weather.[113] Others are considering alternatives such as macadamia, vanilla, ginger, cardamom, turmeric, nutmeg, and cinnamon.[114]

People in the business of coffee are striving to reduce their greenhouse gas emissions. For example, Coopedota, a cooperative in Costa Rica, reduced emissions by 90% by switching to coffee waste products to heat and dry beans. The cooperative also reduced electricity use 40% by using more energy-efficient coffee mills.[115] Starbucks has committed to ten thousand greener stores worldwide to reduce emissions and is helping coffee growers adopt more resilient farming practices.[116] But if we are going to help the millions of coffee farmers stay in business and keep the coffee value chain supplied, we need to do more. Coffee holds a special place on the menu and in the hearts of its many aficionados.

Once again, chocolate is the answer

Carlos J. Pérez has an answer for the world's looming coffee crisis: grow cacao. Pérez, a native Nicaraguan, is the landscape and climate specialist for Solidaridad Network Mesoamerica, which helps growers in Nicaragua, Honduras, Guatemala, and Mexico develop resilience. So many residents in these countries rely on the coffee trade that they risk being deeply affected by climate change.

Solidaridad attracts investments from the private sector or multilateral banks to install cocoa agroforestry systems, mainly in northern Honduras, and joins with industrial partners to integrate small and medium landowners into cocoa production. Why? Cacao trees tolerate the changing climate, there's a growing demand for cocoa, and exporters are buying from small cacao farmers, who can get loans and technical assistance from farming cooperatives. Some members have learned to process chocolate and to diversify further by planting bananas, plantains, and other fruit crops.

(Continued on next page)

(Continued from previous page)

In 2017 Carlos worked with Christian Aid to transition two hundred coffee growers in northern Nicaragua. "We were introducing cacao in areas where farmers were hard hit by coffee rust," says Pérez. Nicaragua ranks eighth in the world for Arabica production, but the drought of 2014 and coffee rust have continued to threaten farmers' livelihoods. For two years, 50% of Nicaragua's coffee crop was lost.

While working with Christian Aid, Pérez and his staff installed twenty simple weather stations on participating Nicaraguan farms. Now growers near those stations can gather rainfall, temperature, and other data, share that with cocoa growers via smart phones, and make better crop management decisions. The system is ramping up capability for short-term forecasts, but Pérez is more interested in what the climate will be like in thirty years. "Are farmers who are shifting from coffee to cocoa more resilient today?" he asks. "What evidence do we have?"

In one project with Solidaridad, he is considering ways to convert degraded pastureland to cacao plantations, making growers eligible for carbon credits. His work supports the conclusions of recent research that every dollar invested in developing resilience will save $2.30 to $13.20 in future humanitarian costs.[a]

---

[a] "The Economics of Early Response and Resilience: Summary of Findings," Government of the United Kingdom, Department of International Development, 17, https://assets.pub lishing.service.gov.uk/media/57a08a0bed915d622c000521/61114_Summary_of_Findings_Final_July_22.pdf.

Summing It Up

The main actors in the finale of the meal, like the foods and beverages that come before them, face an uncertain future. Chocolate, vanilla, dairy products, coffees, and teas are at increasing risk. Many are global commodities, affected by climate change thousands of miles away. But as with the other items on the menu, there are solutions. Farmers are shifting production of some crops to cooler locales and growing shade crops over coffee trees and other plants to reduce heat stress. Breeders are seeking more resilient varieties, and major corporations are making direct investments with producers to ensure the viability of supplies.

For more ways to tackle climate change and save the menu, let's look at what people in the business of food are doing around the world and what we can do individually.

SOLUTIONS

TACKLING CLIMATE CHANGE
AND SAVING THE MENU

Although the prospect of feeding a rapidly growing population under the dark cloud of a changing climate is daunting, solutions do exist, some small, some grand. Many remedies have been shared in previous chapters, and here we suggest more. Given the scale and complexity of the challenges facing our menu, we offer not a comprehensive list but rather key examples of the progress being made and work to be done. We start with what farmers and ranchers are doing, such as climate-smart farming, conserving the soil, diversifying to reduce risks, and adopting new technologies. Those in the business of food are also changing to remain resilient and stay in business. And finally, scientists are developing hardier crops that tolerate the new stresses wrought by climate change.

We close with what we all can do. Become informed, be climate-change literate, speak up about climate change and make it part of the regular dialogue, reduce food waste, switch to a more plant-based diet, and support those who stock the menu—fisher, farmer, and rancher. You can become a force of nature and help bring about the changes we need.

Farmers, Businesses, and Scientists

How They Are Helping

Those who are in the business of growing our food already face a lot of challenges that most people don't appreciate. For example, on any given day, a US farmer might handle issues related to labor, equipment, animal health, a pest infestation, food and worker safety, fuel, seed and fertilizer supplies, and, of course, the weather. Any wrong turn could mean a lost opportunity to plant or harvest that could dramatically affect the bottom line. The farmer might also be at the mercy of a change in consumer preferences or the corporate concentration of power that affects prices for inputs and product sales.[1]

In a crowd of seventy-five employed people in the United States, one is a rancher or farmer—a few feed many.

The number of farmers and ranchers is relatively small in the United States—about 1.3% of the employed population.[2] And about 98% of farms are family farms. Ninety percent of all farms are considered small scale and contribute about 20% to overall production. In contrast, the 3% considered large-scale family farms contribute 46% to overall production.[3] The odds of getting rich at farming are slim. In recent years, half of farm households had a negative farm income, and many relied on off-farm income to make ends meet.[4] Farming is also one of the most hazardous industries, and suicide rates among male farmers in parts of the United States are twice that of the general population.[5] With a changing climate, being a farmer is only going to get more challenging, and many are making the changes needed. As Thor Oechnsner from Oechsner Farms said, "A normal season does not seem like it happens anymore. It's either really dry or really wet. It seems like when we get rain, it's apocalyptic. . . . We got five inches of rain in about one and a half hours, and I had a lot of soil loss. . . . I see the impact for generations."[6]

1. *Practicing climate-smart agriculture.* We touched on several agricultural practices in previous chapters that help to reduce the negative impacts of climate change, including shifting wine grape production north to avoid excessive heat, using shade crops to keep coffee cool as temperatures rise, and using less water in rice paddies to reduce methane emissions. Other solutions include precision irrigation systems, belowground drainage systems, and larger-scale farm equipment so growers can cover more ground during the now-diminishing periods between inclement weather events.

These farming practices are typically considered "climate smart," a widely adopted strategy championed by the Global Alliance for Climate Smart Agriculture and by the North American Climate Smart Agriculture Alliance.[7] The approach takes a holistic view of farm operations to reduce greenhouse gas emissions and increase productivity and income while contributing to food security.[8] Some call this "common sense agriculture" since it makes sense to use the best practices available to stay in business as climate change intensifies. Many farmers have been using "smart" practices for decades.

Climate-smart farming computer programs help northeastern US farmers minimize risk to apples from unanticipated frosts, obtain better crop development and maturity estimates, and use thirty-day forecasts to de-

Farming and ranching made tougher with COVID-19

Farmers in the United States already have a tough job, and the COVID-19 pandemic in 2020 made it much worse. With the closures of schools and restaurants—the main buyers of fresh foods—farmers were forced to dispose of millions of gallons of milk, millions of eggs, and tons of vegetables every day.[a] In addition, a drop in disposable consumer income was expected to result in a decline in consumption of poultry, beef, and pork. Overall, net farm income was projected to drop an estimated $18 billion in 2020. Many impacts were expected to be relatively short-term but dependent on the duration of the pandemic and the post-COVID-19 economic recovery.[b]

---

[a] David Yaffe-Bellany and Michael Corkery, "Dumped Milk, Smashed Eggs, Plowed Vegetables: Food Waste of the Pandemic," *New York Times*, April 11, 2020, https://www.nytimes.com/2020/04/11/business/coronavirus-destroying-food.html.

[b] Food and Agricultural Policy Research Institute, "Baseline Update for US Agricultural Market," June 2020, https://www.fapri.missouri.edu/wp-content/uploads/2020/06/2020-June-Update.pdf.

cide when to irrigate.[9] Many more tools—such as for estimating greenhouse gas emissions and tracking crop health—can be found at the US Climate Resilience Toolkit website.[10] The publicly funded agricultural weather networks in twenty-three states help farmers use these tools optimally and succeed in a riskier world.[11] Last, the Climate Corporation provides farmers with soil, weather, and field data to help optimize yields and reduce climate change risks.[12]

Like climate-smart farming, other strategies such as conservation agriculture, ecological intensification, agroecology, and regenerative agriculture are holistic. Most of these approaches take advantage of natural processes, soil health, and the soil's capacity to sequester carbon. Organic agriculture encompasses many of these same principles.[13] In the United States, consumer demand has driven a double-digit increase in organic products since the 1990s, yet organics still account for only about 4% of total US food sales and about 1% of farmland.[14]

Most US producers today practice what is referred to as conventional or industrial agriculture. This form of farming is often associated with

The one-billion-dollar question for climate-smart agriculture

Rima Al-Azar grew up with a strong sense of social justice, due in part to a father who was a judge. When she was very young, Al-Azar saw a newspaper photograph of the Ethiopian famine and thought, "It's not right that this can happen."

Years later, she earned advanced degrees in the United States and Italy in psychology, international development, and public policy. After gaining field experience in fifty countries, Al-Azar began working in Rome for the United Nations Food and Agriculture Organization (FAO). As team leader on climate-smart agriculture (CSA), she's helping shape the world's response to climate change so that 821 million undernourished people can eat.

The FAO leads the globe in CSA, which has three pillars: increase agricultural production sustainably; adapt agricultural practices to climate change and make them more resilient; and reduce greenhouse gas emissions, when possible. "Agriculture produces 24% of the greenhouse gas emissions in the world," says Al-Azar, "so it's an important sector."

"We estimate that by 2050 the world's population will increase by one-third, and in some countries the population will double." Given how many of the world's people are currently underfed, global agricultural production must increase by 60%. "When people get richer, they tend to want to eat more meat, and animal protein requires more water and inputs to produce. To meet this demand, we have to increase production in a sustainable way."

potential adverse health and environmental impacts, but when farms are sustainably managed, as many are, these risks can be minimized or avoided. Sustainably managed farming is highly efficient and has increased food production over the past fifty years without using much more land. Environmentally friendly conventional farming will be needed to feed a growing global population.[15] In reality, regardless of what category a farm may fall into, many are using similar sound practices. For example, both organic and conventional farms might plant cover crops or rotate crops. Climate-smart agriculture encompasses parts of all farming approaches,

The FAO produces up-to-date information on food and agriculture and disseminates it globally. Her team trains organizations working on climate change and has coordinated the publication of the 570-page *Climate-Smart Agriculture Source Book*.[a] It provides specific information about developing resilience and reducing greenhouse gases in agriculture, forestry, and fisheries. Staff also created CSA e-learning modules in English, French, and Spanish. "This is important," says Al-Azar, "because you can reach a larger number of people worldwide."

Her team members partner with different institutions to promote CSA. They're also starting to work with agencies such as the World Bank and the African Development Bank. "If the multilateral development banks design agricultural projects that are climate smart, then we have a bigger impact than doing a small pilot project here or there," says Al-Azar. In addition, she and her team are working with several organizations to evaluate the impact of CSA programs so they can learn which practices farmers actually adopt long-term and why.

"This is for me the one-billion-dollar question. We are into the business of behavioral change. It's not easy, and we don't know enough yet about what incentives to put into place." She acknowledges agencies have done excellent work on the technical side but concludes, "It's the human being that has to change, not the tree, or the water, or the cow."

---

[a] "Climate Smart Agriculture Sourcebook," Food and Agriculture Organization of the United Nations, 2017, http://www.fao.org/climate-smart-agriculture-sourcebook/en.

since all will face new risks, and we need to develop and implement solutions appropriate for each risk. Essentially, we need to pick and choose the best practices and systems for each farming operation.

2. Practicing soil stewardship. Farmers and ranchers are well aware that maintaining and improving soil health is critical to the viability of the farm now and for future generations. One way they accomplish this is by planting cover crops—generally nonmarketable crops that hold and cover the soil when no commercial crop is being grown, such as during the winter. Cover crops can also add nutrients to the soil and help sequester carbon. Farmers planted cover crops on more than 15 million acres (6 million hectares) in the United States in 2017, 50% more than five years before.[16]

Field to market helps growers become climate resilient

Most people don't understand how hard farmers work to achieve sustainability or how long they've been making that effort. So says Allison Thompson, science and research director at Field to Market: The Alliance for Sustainable Agriculture in Washington, District of Columbia.

Thompson and her colleagues at Field to Market look at how agriculture, energy systems, and the environment interact. They work collectively with farmers, agribusinesses, food and beverage companies selling to consumers, nongovernmental and conservation organizations, and university and governmental partners to build sustainability standards for the entire commodity crop production chain.

When Field to Market started in 2006, Thompson says, farmers told the organization, "We've already made improvements." The statistics backed up their claim. "Absolutely, farmers have done a great job adopting reduced tillage and being more efficient with their resource use," she says. "They've definitely made strides in the past thirty-five years."

But a changing climate means more work to be done. Field to Market aims to provide farmers with useful science-based information on water quality, greenhouse gas emissions, soil health, and more. "So a farmer using our program can look at the metrics and see, 'What is my environmental impact? Is there anything I can I do about that?' That farmer can also report to customers, 'This is what I'm doing to be more sustainable and here are my challenges.'"

Thompson says, "In 2017 we worked with twenty-three hundred individual farmers on 2.8 million acres." They also help food companies reduce their greenhouse gases, largely by keeping in mind such issues as emissions from nitrogen fertilizers as they source crops. "One company might want to source 100% of an ingredient for a product through our program. Another company might want to do a demonstration project with just a small number of producers. Our goal is to get companies engaged in sending a signal to the farmers that this is the industry standard for sustainability."

Field to Market's 2016 national indicators report analyzes statistics on eight sustainability indicators, based mostly on data the US Department of Agriculture collects from farmers.[a] Thompson says these include soil conservation, soil carbon, greenhouse gas emissions, energy use, irrigation efficiency, water quality, land use, and biodiversity.

"When you look at the indicators, sustainability is the same thing to me as climate resilience. What you want to encourage farmers to do is be more resilient because you can't predict for any given region whether it's going to be wetter or drier this year. All you can say with some confidence is that it's going to be more variable. You go talk to these farmers and they're seeing that. In Indiana last summer some farmers had a deluge, and ten miles down the road someone else was dealing with drought conditions. These extremes they're dealing with are really challenging.

"Everything we're talking about with sustainability—being more efficient with resource use, reducing erosion, improving soil quality—all of that will make farmers more resilient to the shifts in weather patterns. We can't promise we're going to make them climate proof, but we're giving them management tools so they're more resilient to the changes that are going to happen."

One method that has significantly increased resilience is soil management, she says. "If you're a conventional farmer, you're tilling every year, maybe losing some soil to erosion, and don't have a way to build that soil back up. You're much more vulnerable when excess rain comes down, washes the soil away, and causes flooding. Or if you have a drought, the soil dries out faster.

"Farmers who are keeping residues in the soil, moving to no tillage to reduce disturbance, and adding cover crops are moving toward continuous coverage of soil to improve soil health and soil carbon. What you see is improvement in how the soil responds both to temperature and water. Soils that have better structure because of this type of management will be able to hold more water and keep that water available longer, so they're less susceptible both to flooding and to drought. Managing soils better helps you be resilient to all sorts of climate stressors. That's where we're seeing a lot of interest and activity amongst various members—focusing on soil health and treating the soil biological community not just as a growth medium but as an ecosystem."

---

[a] *Environmental and Socioeconomic Indicators for Measuring Outcomes of On-Farm Agricultural Production in the United States*, 3rd ed. (Washington, DC: Field to Market: Alliance for Sustainable Agriculture, 2016), https://fieldtomarket.org/national-indicators-report-2016/report-downloads/.

Other options include conservation tillage, which emphasizes less soil disturbance, thus retaining moisture, reducing erosion, and reducing fuel use and labor. Conservation tillage is used on more than 65% of the soybean, corn, and wheat acreage in the United States.[17] Some farmers are turning their attention from annual crops to perennials, including grains. Because these don't need to be replanted each year, the soil doesn't need annual tilling. The Land Institute in Kansas is striving to increase the number of perennial crops and take advantage of their benefits, which include carbon sequestration and improved soil health.[18]

Many producers are also using global positioning systems (GPS) to pinpoint planting and applications of fertilizers and pesticides. For example, because the soil conditions in a field are known in detail, the amount of fertilizer applied is adjusted to the needs of the crop as the equipment passes over the land. This is far more efficient than applying a uniform rate across the entire field.

As previously described, soil—the skin of the planet—and climate change are inextricably linked. Between 2007 and 2016, about 13% of the carbon dioxide in the atmosphere per year came from human activities on the land, including agriculture and forestry.[19] And given that most agricultural soils have lost 30% to 75% of their organic carbon, an opportunity exists for those working the land to address this deficit and help sequester enormous quantities of carbon.[20] Practices that increase soil organic carbon—carbon farming—also improve soil health, benefiting crop productivity. In addition to those mentioned already, improving pastureland, establishing forests on land devoid of trees (afforestation), and integrating trees and shrubs into farming systems (agroforestry) all help sequester carbon in the soil for the long term. These various tactics to increase soil carbon could offset 5% to 15% of global fossil-fuel emissions.[21]

People in the financial, political, and technical sectors are starting to realize the importance of the soil in addressing climate change. One bright spot is the number of multinational companies, such as Diageo, Coca Cola, Fonterra, Olam, Danone, Bayer, and Mars, that now include soil organic carbon as one strategy to increase resilience and profitability within their agricultural value chains.[22] Danone, for example, is promoting regenerative agriculture with its farmers to enhance carbon sequestration. The Danone Ecosystem Fund supports some of these efforts.[23] In addition, Walmart has launched Project Gigaton, an initiative to reduce greenhouse gas emissions by one billion tons by 2030 in partnership with suppliers. Improving soil health at the farm level is one of the options.[24]

3. *Diversifying.* Farmers and ranchers are also minimizing risk by diversifying the crops and animals they produce. Some are planting marketable fruit crops with cacao to ensure a more secure income if market demand and climate change merit it. Planting multiple types of vegetables and at different times of the season is another approach, increasing the odds that if some are lost to severe weather, not all will be. Because 75% of the world's nutrition comes from just twelve crops and five species of livestock, we've put too many "eggs" in one basket.[25] Planting more varied crops helps ensure that even if losses occur to some crops, others remain as income sources. The extent to which farmers can diversify crops will vary by region, existing infrastructure, transportation systems, and markets.

4. *Managing pests and protecting pollinators.* Farmers and pest managers are taking advantage of new monitoring technologies, such as drones and new computer models, that predict climate-induced shifts in new or existing pest threats. At the same time, they're using long-standing management options: planting pest-resistant varieties (including genetically engineered varieties), rotating crops to minimize pests, and preserving beneficial pest-attacking insects. Despite doing most everything to prevent pest infestations and damage, farmers may still need pesticides to protect their crops and prevent unacceptable economic losses. When conditions are right, some crop diseases, such as late blight (known as the "plant destroyer") can raise havoc in tomatoes almost overnight. On conventional farms, the options available for control are generally pesticides. This common-sense approach to managing pests—tapping all options and using pesticide only when truly needed—is called integrated pest management, or IPM.

Farmers can help the pollinators that climate change threatens by planting field margins with flowering plants, limiting tillage outside fields to protect ground-nesting species, and minimizing pesticide risks. In urban and suburban areas residential and community gardens are pollinator "hotspots," and urban planners need to consider enhancing pollinator conservation.[26] Researchers are even testing robotic bees (drones) to pollinate plants, but some argue that we must preserve existing pollinators, which are highly efficient, economical, and exceedingly important to biodiversity.[27]

Building a protein highway with carinata

Carinata, a new oilseed crop

John Oliver has big plans for feeding a world beset by climate change. "North America desperately needs a strategy," he says, "because we're not going to be habitable after 2100 by the way we're going."

Now president of Maple Leaf Bio-Concepts—a consulting firm that helps start-ups succeed—Oliver grew up on a farm in Ontario, Canada, graduated from the Ontario Agricultural College at Guelph, and then promoted agricultural products for Eli Lilly Canada. Later he ran Dow Elanco Canada and helped found the canola industry, developing the cooking oil into Canada's largest cash crop.

With climate change posing a threat to canola, Oliver wants to build a "protein highway" paralleling the United States–Canadian border from the southern Canadian prairies into the Dakotas. In these fertile plains, a three-part cropping system would feature a new cousin to canola—carinata, an African mustard that Agriculture Canada has adapted to North America's growing season.

Carinata produces a small seed full of oil. With a company near Ottawa, Oliver is developing carinata into a biojet fuel that could someday help the aviation industry reduce greenhouse gas emissions. Oliver says livestock can be fed with carinata's protein meal, which is better for dairy cows than canola meal.

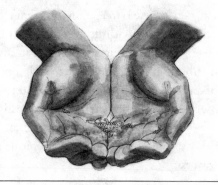

Carinata produces good yields with 20%–40% less water than canola needs and withstands hotter temperatures. Says Oliver, "Our best opportunity is to go where people understand canola culture, but they can't grow canola because the weather's too uncertain." Carinata, like other mustards, opens up the soil and releases chemicals that thwart soil-borne pests and pathogens. Farmers can rotate it with crops that need similar growing conditions, usually wheat and lentils but also garbanzos, soybeans, yellow peas, and other legumes.

"There's been an explosion in the demand for lentils," says Oliver. North American farmers can grow and sell them profitably while reaping benefits from the nitrogen they return to the soil. "Now farmers have a way to grow crops continuously, and it's really opening up opportunities that are good for the environment." Furthermore, growing legumes fits with Canada's Food Guide, which prioritizes plant proteins over red meats.

Right now farmers in Uruguay and Argentina grow the most carinata, which they plant as an off-season rotational crop. Oliver says the demand is so overwhelming that they're not going to be able to satisfy it. Millions of dollars' worth of orders can't be filled because farmers lack the funding to scale up the crop. He estimates carinata could profitably be planted on 8–10 million acres (3.2–4 million hectares), "and we're doing about 150,000."

Yet he's undaunted. When Oliver was a younger man, he broke his spine in a commercial plane crash that was fatal for other passengers. Since then, the paralysis has restricted his mobility but not his dreams. He says, "I believe leadership is largely developing a vision of where you want to go and making it so exciting that everybody wants to be part of getting there."

5. Using less energy and adopting alternative energy sources. Assessing energy use on the farm can identify where farmers can improve efficiency and reduce expenses, perhaps in lighting, crop drying, livestock watering, heating, irrigation, or field and machinery operations. Anaerobic digesters previously discussed convert livestock waste into heat and power. In addition, more than 130,000 farms were producing renewable energy with wind, solar, or geothermal in 2017, double the number in 2012.[28]

6. Investigating new technologies and new kinds of farms. Where feasible, farmers can employ digitization, automation, and artificial intelligence. They can use drones and robots to monitor field conditions, identify pest infestations, and precisely control weeds.[29] Other options include

What food businesses are doing

The Business for Social Responsibility (BSR) group outlines key practices the food, beverage, and agriculture industry are using.[a] Examples include the following:

- Assessing risks along the supply chain and how best to manage them.
- Assessing risks to physical assets, which might involve making emergency response plans.
- Focusing on conservation, including more efficient use of energy and water, in anticipation of increasing costs and decreased availability, respectively.
- Investing in the supplier to ensure resilience to climate change.
- Developing alternative ingredients that are more climate change resilient and have a lower carbon footprint.

Many of these practices are already a standard part of managing business risks. However, many companies may not be prepared for the intensifying—and highly unpredictable—effects of climate change. The BSR identifies several proactive business practices, including enhancing partnerships with communication technology and insurance companies, further diversifying sources (new crops and new production areas), and increasing awareness of how businesses might help contribute to economic, social, and environmental stability for their region's and their company's interests.

---

[a] Joyce Wong and Ryan Schuchard, "Adapting to Climate Change: A Guide for the Food, Beverage, and Agriculture Industry," Business for Social Responsibility, https://www.bsr.org/reports/BSR_Climate_Adaptation_Issue_Brief_Food_Bev_Ag2.pdf.

indoor farming or controlled-environment agriculture, including vertical farms—in which crops are stacked—and hydroponics, by which crops are grown without soil. Such approaches use far less water and nutrients than outdoor production, have minimal pest infestations, and allow year-round agriculture in urban areas and other locations close to consumers. These technologies produce many more pounds of product per unit area than field-grown crops, but they are expensive and generally come with a high carbon footprint.[30] However, if costs and the carbon footprint can be reduced, indoor farming may play a larger role in the future as climate change further afflicts outdoor production. Locating in urban areas would have the advantage of very low transport distances.

What Scientists Are Doing

To keep up with the increasing risks facing agriculture and the food system, we need research, and a lot of it. The *Fourth National Climate Assessment* identified areas that need to be studied, including how climate change affects pests; how increasing carbon dioxide, changes in water availability, and temperature interact and affect crops in the field; and how climate resilience across many crop types can be improved.[31] This is not going to happen without time and money—and we are running out of both. Unfortunately, federal funding for agricultural research and development has been declining. This research often addresses practical near-term needs as well as longer-term solutions—for the public good and available to everyone. In contrast, private funding has been increasing, but it generally focuses on commercialization of products and technologies, with an expectation of financial return. Funding from individual state governments, private foundations, and commodity or producer groups also helps but for the last two may come with a need for a match with government funds. In 2015, federal funds for agriculture research and development totaled $4.5 billion.[32] To put this in perspective, people in the United States spent over $43 billion on video games, accessories, and hardware in 2018.[33]

Public investments in innovation and technology advances are the main drivers of agricultural economic growth in the United States.[34] In addition, public funds typically benefit agriculture more broadly than private investments. Public investment is important and getting more important daily.

Public dollars support the US Department of Agriculture's (USDA) Agricultural Research Service, which studies adaptation to climate change and mitigation strategies for agriculture, and the Natural Resources Conservation Service, which develops region-specific technologies focused on climate change adaptation.[35] The USDA's National Institute for Food and Agriculture supports research and outreach critical to agriculture and food systems.[36] The ten USDA Climate Hubs foster collaboration within USDA and with regional partners to develop and deliver science-based information to agricultural and natural resource managers.[37] The six National

Oceanic and Atmospheric Administration Regional Climate Centers monitor and study regional weather and climate and share that information widely.[38] These help policy makers, planners, businesses, and farmers. Experts at the US Global Change Research Program study how climate change and other forces are changing global systems and how to respond and plan accordingly.[39] They periodically publish the *National Climate Assessment*, a science-based report on climate change and its US impacts. The land grant universities, which generate "knowledge with public purpose," are in every state and also play a critical role. These institutions, which encompass research, education, and outreach through Cooperative Extension, are well placed to address local and regional impacts of climate change. To succeed, however, the system needs adequate funding and must address climate change along with other issues facing its clientele.[40]

More Resilient Crops

Crop plants—the basis of the food chain—must become more resilient. We have shared several examples of how scientists are developing barley and hops (for beer) that tolerate salt, heat, and drought; heat-tolerant potatoes; and salt-tolerant rice. Farmers and many others in the business of food benefit from these developments. Here we provide an overview of the ways more resilient crops and animals are developed through selection and breeding— an approach of profound importance given the threat of climate change.

Since the beginning of agriculture, humans have been selecting highly productive plants that resist pests or survive floods and droughts— essentially those with the right genes. For example, very early farmers saved seeds from the best grains year after year, essentially practicing long-term genetic modification. Today, crop plant improvement usually involves the purposeful transfer of pollen from a male parent to a female parent—for example, crossing a drought-tolerant variety with a high-yielding one to achieve a new type with both characteristics.

Plant breeders might repeat this process for years before they achieve the desired results, and then seed farms make seeds of the improved variety available to farms. Developing fruit or nut trees takes even longer because trees take time to mature and grow fruit and seed. Most of us have eaten traditionally crossbred crops, which have been part of the menu for at least one hundred years.

Traditional plant breeding continues around the globe, but new methods can directly transfer into a plant one or more specific traits from the same species, a closely related one, or a totally different one. This genetic modification or genetic engineering results in "transgenic" plants, commonly referred to as GMOs (genetically modified organisms). Although genetic engineering of crops replaces labor-intensive cross-pollination, it can still take several years from the initial crop modification to commercial release.

Scientists have had the ability to edit a plant's genetic makeup for a few decades, but the technology has recently advanced thanks to "clustered regularly interspaced short palindromic repeats," or CRISPR. Scientists using CRISPR can delete, replace, or tweak genes, or insert novel genes from a different organism. It was considered science's breakthrough of the year by *Science Magazine* in 2015.[41] As long as there are no novel genes inserted into the edited plant, the USDA considers it equivalent to a plant improved through traditional breeding methods.[42] It is not considered a "GMO." The CRISPR technology has the capacity to improve crop yield, quality, resistance to insects and diseases, tolerance to heat stress, and much more.[43]

Genetically engineered crops that help manage pests, tolerate weed killers, and increase nutritional qualities and shelf life are available, and some have been in use for decades. One of the earliest was *Bt* corn. Short for the bacterium *Bacillus thuringiensis*, *Bt* occurs naturally in the soil and has a long history as a pest-selective insecticide applied on both organic and conventional farms. A protein from this bacterium in *Bt* crops turns

Gene editing is precise—it's like removing a single gene from a corn plant that has thirty-two thousand genes or one key from a piano with the same number of keys.

into a toxin when an insect pest ingests it. Today several versions of *Bt* crops target particular insect pests, and in 2018 over 80% of US cotton and field corn were *Bt* types.[44]

Several crops, including soybeans, cotton, corn, sugar beets, alfalfa, and canola have been engineered to tolerate the herbicides that normally would kill them. For example, glyphosate, the main active ingredient in Roundup, was designed to control all types of plants except the herbicide-tolerant crop. As with *Bt* crops, scientists obtained the gene that provides the crop tolerance to glyphosate from naturally occurring soil bacteria. About 90% of the soybean, cotton, and field corn acres grown in the United States are herbicide tolerant.[45]

Researchers are pursuing some potential game changers, such as increasing plants' photosynthesis rates. Initial studies show plants genetically engineered for a higher photosynthesis rate are about 40% more productive than the wild type.[46] Scientists have also proposed that, through selection and breeding, some crops could turn increased carbon dioxide levels into higher yields.[47] Finally, gene editing technologies give us the potential to domesticate wild species and help diversify our food supply.[48]

People have raised economic, environmental, and human health concerns about genetically engineered crops. However, in 2016 the National Academies of Sciences, Engineering, and Medicine indicated that US farmers generally profited from using genetically engineered crops such as field corn, cotton, and soybeans. They also found environmental benefits from the reduced use of insecticides and reduced yield losses from the use of *Bt* varieties. In addition, widespread planting of some *Bt* crops resulted in region-wide suppression of specific pests, benefiting even crops without the *Bt* trait. Insect biodiversity was also higher in *Bt* crops than in non-*Bt* crops for which synthetic insecticides were used. The authors noted that farmers planting herbicide-resistant varieties did not need to cultivate as much as farmers who planted conventional crops, thus reducing the potential for soil erosion. Although insecticide use declined overall, use of the herbicide glyphosate increased with the adoption of herbicide-resistant crops while use of other herbicides decreased. Most importantly, the assessment also concluded that foods using genetically engineered crops pose no more health risks than those using conventionally bred crops.[49]

In 2016, twenty-six countries planted transgenic crops on 460 million acres (185 million hectares).[50] As the use of transgenic crops expands, each should be evaluated for risks. The National Academies report recommends safety testing of new varieties, whether traditionally bred or genetically engineered, if they have novel characteristics with potential hazards. In addition, the oversight should be participatory and transparent.[51] And though these crops have many benefits, repeated use can reduce their effectiveness against pests. Just as the bacteria, fungi, and parasites that cause infectious human diseases have become resistant to widely used drugs, certain weeds are now resistant to glyphosate, and some insects are resistant to *Bt*.[52] Gene stacking—combining two or more traits into one plant—can overcome the development of resistance by a pest. For example, plants with multiple types of *Bt* have several ways to control a pest, not just one.[53]

Despite the benefits and relatively few risks, some countries have banned or limited the use of transgenic crops.[54] To help inform US consumers, the USDA in 2019 required labeling, starting in 2022, of all food products intended for human consumption that contain genetically engineered ingredients.[55]

The benefits of genetic engineering extend well beyond plants. Genetically engineered mice are important in studying causes of and cures for some human diseases.[56] The insulin used worldwide for type 1 diabetes is produced using genetically engineered bacteria and yeast.[57] Scientists see many more opportunities for genetic engineering in coming years, including treatment and diagnosis of diseases and production of vaccines and other drugs.

Summing It Up

It's all hands on deck. Farmers need to adopt climate-smart and new precision agriculture technologies and diversify their crops. Scientists need to develop more resilient crops, and food businesses need to assess risks along their supply chains. Public support for research and development must increase to meet the onslaught of intensifying risks. Public–private partnerships will be critical, and together we can make progress in saving the menu.

What We Can Do

Farmers, food businesses, and scientists are doing their part. Now let's turn to what each of us can do. Knowing that climate change poses increasing risks to the foods and beverages we need and love, we must adapt and cut greenhouse gas emissions in every way possible. Those of us living in rich nations such as the United States are the biggest contributors to climate change and are best positioned to tackle it.

1. Become climate change literate. This book has helped you understand how and why the climate is changing and the impacts, especially on the menu, but read on to learn more and challenge yourself to stay up-to-date on how our world is rapidly changing. Informed and climate-change-literate people

- understand how they influence the climate and how it influences them and society;
- understand the essential principles of Earth's climate system;
- know how to assess scientifically credible information about climate;
- communicate about climate and climate change in a meaningful way; and
- make informed, responsible decisions about actions affecting the climate.[1]

Once you are climate change literate, help others become literate as well. When people understand that their actions affect climate change, they also see how their actions can help reverse it. For a climate change primer, see the National Oceanic and Atmospheric Administration's *Climate Literacy: The Essential Principles of Climate Science* (https://downloads .globalchange.gov/Literacy/climate_literacy_highres_english.pdf).

2. Start talking about climate change. Katharine Hayhoe, the rock star of climate change communication, recommends making these discussions a common occurrence—a social norm.[2] In 2019, surveys indicated that 36% of people in the United States occasionally talked about global warming and 64% rarely or never talked about it.[3] You can help change that. Bring up climate change where you work, worship, attend school, shop, live, and anywhere else. Research suggests that talking to your friends and family leads to deeper engagement.[4] Speak up. Raise your voice. Here are a few suggestions from the Nature Conservancy and elsewhere:[5]

- Meet people where they are, listen to what they say, respond to their comments, and be patient.
- Feelings are more important than facts. Make your comments relevant by connecting climate change to what you see where you live and work. Express your personal concern for kids, grandkids, and their future.
- Make it a conversation, not an argument. If you keep the connection, you can potentially pick up the dialogue again in the future.
- Talk about solutions in your personal life and workplace.
- Avoid overwhelming your listener with facts, and don't use fear.
- Despite your best efforts, it's not going to work every time, but give it a try.

Everyone eats. Talking about food can transcend political, sociological, and cultural barriers and build bridges across divisiveness. Having read this book, you are now prepared to tell a new story about climate change. Use your favorites—coffee, chocolate, potatoes, or your special comfort food—to convey the urgent need for action. By saving and preserving the breakfast, lunch, and dinner menu, we may be able to save humanity.

3. Move to a more plant-based diet. The greatest impact we can have with our food choices is to transition to meals that are more plant-based. This change is not for everyone, especially in regions of the world where meat is a critically important source of protein and nutrients. The human race is unlikely to stop eating animal products completely, but doing so would have impressive benefits—a 76% reduction in food's land use and about a 50% reduction in food's greenhouse gas emissions worldwide. In the United States, where people eat three times more meat than the global average, food-related emissions would be reduced up to 73%. Even a 50%

A simple exercise

Almost all our foods and beverages—from A to Z—are changing, and here's a simple exercise to drive home the point. Search the internet using the words "climate change" followed by any one of the items listed below and see what you discover. Then do the same with your favorite foods. You will likely find many climate change stories, some challenges, some opportunities, and some surprises. Be brave and try this with others at your next meal. It's guaranteed to get you talking about climate change and food!

abalone	eggplant	pasta
açaí palm	figs	peas
asparagus	garlic	pork
Assamese black tea	ginseng	quinoa
bananas	goose	radish
blackberries	herbs	spices
bluefin tuna	huckleberries	steak
Brazil nuts	juniper	St. John's wort
cardamom	kelp	thyme
caviar	kiwi	trout
citrus	mango	venison
crayfish	mushroom	vetiver
dragon fruit	nutmeg	yams
Dungeness crab	oregano	Yunnan pu'erh tea
durian	palm oil	zucchini

reduction in global consumption of animal products would have a huge positive impact.[6]

Achieving this goal would be tough given the cultural, economic, nutritional, and even evolutionary connections humans have to animal products in their diets. In the United States, for example, only 5% of people consider themselves vegetarian,[7] and in several European countries, it's 6% to 10%.[8] The Lancet Commission reported that a diet with just a small quantity of red meat wouldn't exceed the planet's available resources. Such a diet could improve your health and the planet's—a very doable and worthy goal.[9] In the chapter "The Main Course," we mentioned some menu options with lower carbon footprints than conventionally produced beef, among them plant-based meat alternatives.

4. Reduce food waste. Food loss and waste is highly underrated as a global problem and as a contributor to climate change. Getting that food to the table—producing, transporting, processing, and storing it—releases greenhouse gases and when food decays in a landfill it emits more gases. The United Nations' Food and Agriculture Organization reported in 2013 that if food waste were a country it would be the third largest greenhouse gas emitter globally. About one-third of the global food supply is never eaten.[10] If we saved only one-fourth of this waste, it would feed 870 million people annually. In North America, about 40% of food wastage occurs at the consumption stage mainly due to somewhat subjective "sell-by date" labeling and because consumers like produce without blemishes.[11] Do we need to be so fussy? Food with blemishes is just as healthy and flavorful.

The US Environmental Protection Agency suggests buying only what we need, storing fruits and vegetables for maximum freshness, using leftovers and old ingredients before they spoil, and understanding food product dating.[12] They also have a "Too Good to Waste Toolkit" that helps determine how much food we waste at home and how to waste less.[13] To make progress globally, waste in the entire food supply chain needs attention, including consumer awareness and consideration for recycling waste back into the system.[14] It may also require creativity, like the Brits making beer from some of the twenty-four million slices of bread discarded every day.[15]

5. Consider your entire carbon footprint. This book has focused on food production, but let's keep things in perspective. In the United States, the agricultural sector is responsible for about 9% of the country's greenhouse gases, but that means 91% comes from somewhere other than agriculture—transportation, electricity production, industry, and the residential and commercial sectors.[16] To solve the climate change crisis we need to reduce emissions across all sectors—now. Fly, drive, light, heat, and cool less and consume less stuff. You can assess your carbon footprint with a calculator like the one the Global Footprint Network provides.[17] You can also review Drawdown, which prioritizes options to reduce greenhouse gases at a global scale.[18] Determine where you can play a role and the bigger the better.

6. Appreciate and support the people who supply the menu. As described in the previous chapter, farming is hard work, and Dwight Eisenhower drove that point home when he said "farming looks mighty easy when your plow is a pencil, and you're a thousand miles from the corn field."[19] Consider pausing before your next meal to appreciate where your food comes from. Thank the people who make it possible, whether it's a family harvesting cacao in western Africa, a boat captain catching cod in the Bering Sea, or a vegetable farmer 30 mi. (48 km) away. To support nearby producers, buy local or join a CSA (Community Supported Agriculture), then expand your efforts to help the broad community of people involved in food, especially those on the front lines. We now offer some ideas on how to make progress.

7. Be an activist and help create the change we need now. Having read this book, you can share the food and climate change story with policy makers, members of the business community, and many others. Engage elected officials—rural and urban—on the topic of food and climate change and start the dialogue to see how aware they are of the local and global impacts. You may face partisan barriers but start the dialogue with those in positions of influence.[20] Raise this critically important topic to a new level and start with the increasing risks to the foods we all love and need. Add to this the potential economic impacts: food is a big business and employs millions of people.

Challenge policy makers to support programs that help people who produce our food to stay in business as well as minimize their impacts. These programs should extend globally because food insecurity is likely to lead to increasing conflict and social unrest with worldwide implications.[21] Ask elected officials to support the science that is needed now more than ever. Along with your actions, engage others to work together to help those who produce our food and keep the menu stocked.

A Call to Be Courageous

To tackle the grand challenge of climate change will require every one of us. Experts indicate that the earth's climate could actually be stabilized by 2050 if the necessary transformative technological and social changes are set in motion and spread quickly. These are much like climate change tipping points, unstoppable, irreversible, but in this case, positive. They include incentivizing a massive shift away from fossil fuels, sharing information on sources of emissions, helping people realize the growing risks, shifting from

> # Remember what we can do!
>
> 1. Become climate change literate and help others learn.
> 2. Start talking about climate change.
> 3. Adopt a more plant-based diet.
> 4. Reduce food waste.
> 5. Consider your entire carbon footprint.
> 6. Appreciate and support the people who supply the menu.
> 7. Be an activist and help create the change we need now.

a grassroots movement to a global network, changing lifestyles such as diets, and educating others about climate change in new and improved ways.[22]

Given that food is universal, has a huge carbon footprint, is at risk, and is intimately tied to cultures and lifestyles, it could be a social tipping point. Imagine climate change being discussed over food and drink. If consumers plus those in the business of food—producers, chefs, restaurateurs, retailers, manufactures, and all the others—join forces and raise their voices it could be transformative.

We hope that *Our Changing Menu* will help start a great awakening for all, but especially for those not yet convinced. It is now your turn. Help build momentum and inspire others to join in. Be courageous.

NOTES

Preface

1. Jonathon P. Schuldt, Danielle L. Eiseman, and Michael P. Hoffmann, "Public Concern about Climate Change Impacts on Food Choices: The Interplay of Knowledge and Politics," *Agriculture and Human Values* (January 29, 2020):6, https://doi.org/10.1007/s10460-020-10019-7.

Our Food Supply

1. Sarah A. Low et al., "Trends in U.S. Local and Regional Food Systems: A Report to Congress," US Department of Agriculture (USDA) Economic Research Service AP-068 (January 2015), 2. https://www.ers.usda.gov/publications/pub-details/?pubid=42807.
2. USDA Economic Research Service, "Americans Consume Mostly U.S.-Made Food, Produce," *Western Livestock Journal*, June 11, 2018, https://www.wlj.net/top_headlines/americans-consume-mostly-u-s--made-food-produce/article_a76f95f0-5857-11e8-8922-47f84163101f.html; "Agricultural Trade," USDA Economic Research Service, last updated August 20, 2019, https://www.ers.usda.gov/data-products/ag-and-food-statistics-charting-the-essentials/agricultural-trade/.
3. "U.S. Food Imports," USDA Economic Research Service, last updated August 20, 2019, https://www.ers.usda.gov/data-products/us-food-imports/.
4. Jessica A. Gephart, Halley E. Froehlich, and Trevor A. Branch, "Opinion: To Create Sustainable Seafood Industries, the United States Needs a Better Accounting of Imports and Exports," *Proceedings of the National Academy of Sciences* 116, no. 19 (May 7, 2019):9142–46, https://doi.org/10.1073/pnas.1905650116.
5. Aleda V. Roth et al., "Unraveling the Food Supply Chain: Strategic Insights from China and the 2007 Recalls," *Journal of Supply Chain Management* 44, no. 1 (January 2008):24, https://doi.org/10.1111/j.1745-493X.2008.00043.x.
6. Brian Halweil and Tom Prugh, *Home Grown: The Case for Local Food in a Global Market*, Worldwatch Paper 163 (Washington, DC: Worldwatch Institute, 2002), 18.
7. Sophia Murphy, David Burch, and Jennifer Clapp, "Cereal Secrets: The World's Largest Grain Traders and Global Agriculture," Oxfam Research Reports

(August 3, 2012), 3, https://www.oxfam.org/en/research/cereal-secrets-worlds-largest
-grain-traders-and-global-agriculture.

8. *Annual Review 2018*, Nestlé, https://www.nestle.com/sites/default/files/asset
-library/documents/library/documents/annual_reports/2018-annual-review-en
.pdf, 47, 55.

9. Lutz Goedde, Maya Horii, and Subuk Sanghvi, "Pursuing the Global Opportunity in Food and Agribusiness," McKinsey & Company, accessed March 22, 2019,
https://www.mckinsey.com/industries/chemicals/our-insights/pursuing-the
-global-opportunity-in-food-and-agribusiness.

10. "Ag and Food Sectors and the Economy," USDA Economic Research Service, last
updated September 20, 2019, https://www.ers.usda.gov/data-products/ag-and-food
-statistics-charting-the-essentials/ag-and-food-sectors-and-the-economy/.

11. Polly J. Ericksen, "Conceptualizing Food Systems for Global Environmental
Change Research," *Global Environmental Change* 18, no. 1 (February 2008):236,
https://doi.org/10.1016/j.gloenvcha.2007.09.002.

12. Hannah Ritchie, "Yields vs. Land Use: How the Green Revolution Enabled Us to
Feed a Growing Population," Our World in Data, August 22, 2017, https://
ourworldindata.org/yields-vs-land-use-how-has-the-world-produced-enough
-food-for-a-growing-population.

13. H. Charles J. Godfray et al., "Food Security: The Challenge of Feeding 9 Billion
People," *Science* 327, no. 5967 (February 12, 2010):812–18, https://doi.org/10.1126
/science.1185383.

Our Changing Climate

1. "Peeling Back the Layers of the Atmosphere," National Oceanic and Atmospheric
Administration (NOAA) National Environmental Satellite, Data, and Information
Service, February 22, 2016, https://www.nesdis.noaa.gov/content/peeling-back
-layers-atmosphere.

2. James E. Hansen and Makiko Sato, "Paleoclimate Implications for Human-Made
Climate Change," in *Climate Change: Inferences from Paleoclimate and Regional
Aspects*, ed. André Berger, Fedor Mesinger, and Djordje Sijacki (Vienna: Springer,
2012), 37, https://doi.org/10.1007/978-3-7091-0973-1_2.

3. K. J. Hayhoe et al., "Climate Models, Scenarios, and Projections," in *Climate Science Special Report: Fourth National Climate Assessment*, vol. 1, ed. D. J. Wuebbles et al. (Washington, DC: US Global Change Research Program, 2017), 138,
https://doi.org/10.7930/J0WH2N54.

4. "Cars Produced in the World," Worldometers, accessed September 9, 2019, https://
www.worldometers.info/cars/; Matthew N. Smith, "The Number of Cars Worldwide Is Set to Double by 2040," World Economic Forum, April 22, 2016,
https://www.weforum.org/agenda/2016/04/the-number-of-cars-worldwide-is-set
-to-double-by-2040/.

5. "Airline Industry Worldwide: Number of Flights 2019," Statista, release date December 2019, https://www.statista.com/statistics/564769/airline-industry-number -of-flights/.

6. "Sources of Greenhouse Gas Emissions," US Environmental Protection Agency, December 29, 2015, https://www.epa.gov/ghgemissions/sources-greenhouse-gas -emissions.

7. Dieter Lüthi et al., "High-Resolution Carbon Dioxide Concentration Record 650,000–800,000 Years before Present," *Nature* 453 (May 2008):380, https://doi.org /10.1038/nature06949.

8. Chloé Farand, "2019 Second Warmest Year on Record, Ends Hottest Decade Yet, Says EU Observatory," Climate Home News, January 8, 2020, https://www .climatechangenews.com/2020/01/08/2019-second-warmest-year-record-ends -hottest-decade-yet-says-eu-observatory/.

9. Myrna H. P. Hall and Daniel B. Fagre, "Modeled Climate-Induced Glacier Change in Glacier National Park, 1850–2100," abstract, *BioScience* 53, no. 2 (February, 2003):131, https://doi.org/10.1641/0006-3568(2003)053[0131:MCIGCI]2.0.CO;2.

10. Phillip D. A. Kraaijenbrink et al., "Impact of a Global Temperature Rise of 1.5 Degrees Celsius on Asia's Glaciers," *Nature* 549 (September 2017):257–60, https:// doi.org/10.1038/nature23878.

11. "Arctic Sea Ice News and Analysis," National Snow and Ice Data Center, September 23, 2019, http://nsidc.org/arcticseaicenews/2019/09/.

12. "Quick Facts on Ice Sheets," National Snow and Ice Data Center, accessed July 22, 2019, https://nsidc.org/cryosphere/quickfacts/icesheets.html.

13. "Ice Sheets," National Aeronautics and Space Administration, accessed July 23, 2019, https://climate.nasa.gov/vital-signs/ice-sheets.

14. W. V. Sweet et al., "Sea Level Rise," in *Climate Science Special Report*, ed. Wuebbles et al., 333, https://doi.org/10.7930/J0VM49F2.

15. Sweet et al., 333.

16. Vickie Machado, "Saltwater Intrusion," WUFT News, accessed February 11, 2020, https://www.wuft.org/specials/water/saltwater-intrusion/; Tran Thi Nhung et al., "Salt Intrusion Adaptation Measures for Sustainable Agricultural Development under Climate Change Effects: A Case of Ca Mau Peninsula, Vietnam," *Climate Risk Management* 23 (January 1, 2019):88–100, https://doi.org/10.1016/j.crm.2018 .12.002.

17. "Ocean Acidification," NOAA, November 2013, https://www.noaa.gov/education /resource-collections/ocean-coasts-education-resources/ocean-acidification.

18. D. J. Wuebbles et al., "Executive Summary," in *Climate Science Special Report*, ed. Wuebbles et al., https://doi.org/10.7930/J0DJ5CTG.

19. Barbara J. Bentz et al., "Climate Change and Bark Beetles of the Western United States and Canada: Direct and Indirect Effects," *BioScience* 60, no. 8 (September 1, 2010):602–13, https://doi.org/10.1525/bio.2010.60.8.6.

20. Patrick Greenfield, "This Is Not How Sequoias Die. It's Supposed to Stand for Another 500 Years," *Guardian*, January 18, 2020, https://www.theguardian.com

/environment/2020/jan/18/this-is-not-how-sequoias-die-its-supposed-to-stand
-for-another-500-years-aoe.

21. "Survival by Degrees: 389 Bird Species on the Brink," National Audubon Society, accessed March 11, 2020, https://www.audubon.org/climate/survivalbydegrees.

22. Sandra Díaz et al., "Summary for Policymakers of the Global Assessment Report on Biodiversity and Ecosystem Services of the Intergovernmental Science-Policy Platform on Biodiversity and Ecosystem Services," Intergovernmental Science-Policy Platform on Biodiversity and Ecosystem Services (2019), 4, https://ipbes .net/sites/default/files/ipbes_7_10_add.1_en_1.pdf.

23. Zeke Hausfather et al., "Evaluating the Performance of Past Climate Model Projections," *Geophysical Research Letters* 47, no. 1 (January 2020):e2019GL085378, https://doi.org/10.1029/2019GL085378.

24. V. Masson-Delmotte et al., "Summary for Policymakers," in *Special Report: Global Warming of 1.5°C*, Intergovernmental Panel on Climate Change, https://www.ipcc .ch/sr15/chapter/spm/.

25. Toby R. Ault et al., "Relative Impacts of Mitigation, Temperature, and Precipitation on 21st-Century Megadrought Risk in the American Southwest," *Science Advances* 2, no. 10 (October 2016):1–8, https://doi.org/10.1126/sciadv.1600873.

26. A. Park Williams et al., "Large Contribution from Anthropogenic Warming to an Emerging North American Megadrought," *Science* 368, no. 6488 (April 17, 2020):314–18, https://doi.org/10.1126/science.aaz9600.

27. Timothy M. Lenton et al., "Climate Tipping Points—Too Risky to Bet Against," *Nature* 575, no. 7784 (November 2019):592–95, https://doi.org/10.1038/d41586-019 -03595-0.

28. Masson-Delmotte et al., "Summary for Policymakers," 4.

29. Lenton et al., "Climate Tipping Points," 595.

30. Abraham Lincoln, "Second State of the Union Address," December 1, 1862, https://en.wikisource.org/wiki/Abraham_Lincoln%27s_Second_State_of_the _Union_Address.

Climate Change

1. Mark D. Risser and Michael F. Wehner, "Attributable Human-Induced Changes in the Likelihood and Magnitude of the Observed Extreme Precipitation during Hurricane Harvey," *Geophysical Research Letters* 44, no. 24 (2017):12,457, https:// doi.org/10.1002/2017GL075888.

2. Kevin Trenberth, "Changes in Precipitation with Climate Change," abstract, *Climate Research* 47, no. 1 (March 31, 2011):123, https://doi.org/10.3354/cr00953.

3. D. R. Easterling et al., "Precipitation Change in the United States," in *Climate Science Special Report: Fourth National Climate Assessment*, vol. 1, ed. D. J. Wuebbles et al. (Washington, DC: US Global Change Research Program, 2017), 207, https://doi.org/10.7930/J0H993CC.

4. David W. Wolfe et al., "Unique Challenges and Opportunities for Northeastern US Crop Production in a Changing Climate," *Climatic Change* 146, no. 1 (January 1, 2018):240, https://doi.org/10.1007/s10584-017-2109-7.

5. P. Gowda et al., "Agriculture and Rural Communities," in *Impacts, Risks, and Adaptation in the United States: Fourth National Climate Assessment,* vol. 2, ed. D. R. Reidmiller et al. (Washington, DC: US Global Change Research Program, 2018), 398, https://doi.org/10.7930/NCA4.2018.CH10.

6. "Groundwater Resources around the World Could Be Depleted by 2050s," American Geophysical Union, December 15, 2016, https://phys.org/news/2016-12 -groundwater-resources-world-depleted-2050s.html.

7. Laura Parker, "What Happens to the U.S. Midwest When the Water's Gone?," *National Geographic*, August 2, 2019, https://www.nationalgeographic.com/magazine /2016/08/vanishing-midwest-ogallala-aquifer-drought/.

8. Brian D. Smerdon, "A Synopsis of Climate Change Effects on Groundwater Recharge," abstract, *Journal of Hydrology* 555 (December 1, 2017):125, https://doi .org/10.1016/j.jhydrol.2017.09.047.

9. Brian Barth, "Farmers Reeling from Record Year of Wildfires," *Modern Farmer*, September 10, 2018, https://modernfarmer.com/2018/09/farmers-reeling-from -record-year-of-wildfires/; M. F. Wehner et al., "Droughts, Floods, and Wildfires," in *Climate Science Special Report*, ed. Wuebbles et al., 231, https://doi.org/10.7930 /J0CJ8BNN.

10. Wolfe et al., "Unique Challenges and Opportunities," 234.

11. Trenberth, "Changes in Precipitation," 123; Noah Knowles, "Trends in Snow Cover and Related Quantities at Weather Stations in the Conterminous United States," *Journal of Climate* 28, no. 19 (July 22, 2015):7518–28, https://doi.org/10.1175/JCLI -D-15-0051.1.

12. "Value of U.S. Agricultural Imports from Peru from 1990 to 2018," Statista, released 2019, https://www.statista.com/statistics/221625/value-of-us-agricultural -imports-from-peru-since-1990/; "Value of U.S. Agricultural Imports from Chile from 1990 to 2018," Statista, released 2019, https://www.statista.com/statistics /221532/value-of-us-agricultural-imports-from-chile-since-1990/.

13. Mathias Vuille et al., "Rapid Decline of Snow and Ice in the Tropical Andes— Impacts, Uncertainties and Challenges Ahead," abstract, *Earth-Science Reviews* 176 (January 1, 2018):195, https://doi.org/10.1016/j.earscirev.2017.09.019.

14. Eklabya Sharma et al., "Introduction to the Hindu Kush Himalaya Assessment," in *The Hindu Kush Himalaya Assessment: Mountains, Climate Change, Sustainability and People*, ed. Philippus Wester et al. (Cham, Switzerland: Springer, 2019), 4, https://doi.org/10.1007/978-3-319-92288-1_1.

15. Julian C. Brimelow, William R. Burrows, and John M. Hanesiak, "The Changing Hail Threat over North America in Response to Anthropogenic Climate Change," *Nature Climate Change* 7, no. 7 (July 2017):516–22, https://doi.org/10.1038 /nclimate3321.

16. W. J. W. Botzen, L. M. Bouwer, and J. C. J. M. van den Bergh, "Climate Change and Hailstorm Damage: Empirical Evidence and Implications for Agriculture and Insurance," *Resource and Energy Economics* 32, no. 3 (August 1, 2010):342, https://doi.org/10.1016/j.reseneeco.2009.10.004.

17. "Guidance for Industry: Evaluating the Safety of Flood-Affected Food Crops for Human Consumption," US Food and Drug Administration, October 2011, http://www.fda.gov/regulatory-information/search-fda-guidance-documents/guidance-industry-evaluating-safety-flood-affected-food-crops-human-consumption.

18. Hans W. Paerl and Jef Huisman, "Climate Change: A Catalyst for Global Expansion of Harmful Cyanobacterial Blooms," *Environmental Microbiology Reports* 1, no. 1 (2009):27, https://doi.org/10.1111/j.1758-2229.2008.00004.x.

19. Jennifer C. Phillips et al., "The Potential for CO_2-Induced Acidification in Freshwater: A Great Lakes Case Study," abstract, *Oceanography* 28, no. 2 (October 2015):136, https://doi.org/10.5670/oceanog.2015.37.

20. Annie Sneed, "Ask the Experts: Does Rising CO_2 Benefit Plants?," *Scientific American*, January 23, 2016, https://www.scientificamerican.com/article/ask-the-experts-does-rising-co2-benefit-plants1/.

21. Chunwu Zhu et al., "Effect of Nitrogen Supply on Carbon Dioxide-Induced Changes in Competition between Rice and Barnyardgrass (*Echinochloa Crus-Galli*)," abstract, *Weed Science* 56, no. 1 (February 2008):66, https://doi.org/10.1614/WS-07-088.1.

22. João Paulo Refatti et al., "High [CO_2] and Temperature Increase Resistance to Cyhalofop-Butyl in Multiple-Resistant *Echinochloa Colona*," *Frontiers in Plant Science* 10 (2019):1–11, https://doi.org/10.3389/fpls.2019.00529.

23. Jason G. Hamilton et al., "Anthropogenic Changes in Tropospheric Composition Increase Susceptibility of Soybean to Insect Herbivory," abstract, *Environmental Entomology* 34, no. 2 (April 1, 2005):479, https://doi.org/10.1603/0046-225X-34.2.479.

24. Samuel S. Myers et al., "Increasing CO_2 Threatens Human Nutrition," *Nature* 510, no. 7503 (June 2014):139–42, https://doi.org/10.1038/nature13179.

25. Chunwu Zhu et al., "Carbon Dioxide (CO_2) Levels This Century Will Alter the Protein, Micronutrients, and Vitamin Content of Rice Grains with Potential Health Consequences for the Poorest Rice-Dependent Countries," *Science Advances* 4, no. 5 (May 23, 2018):2, https://doi.org/10.1126/sciadv.aaq1012.

26. Matthew R. Smith and Samuel S. Myers, "Impact of Anthropogenic CO_2 Emissions on Global Human Nutrition," abstract, *Nature Climate Change* 8, no. 9 (September 2018):834, https://doi.org/10.1038/s41558-018-0253-3.

27. Lewis H. Ziska et al., "Rising Atmospheric CO_2 is Reducing the Protein Concentration of a Floral Pollen Source Essential for North American Bees," *Proceedings of the Royal Society B: Biological Sciences* 283, no. 1828 (April 13, 2016):20160414, 3, https://doi.org/10.1098/rspb.2016.0414.

28. Deke Arndt, "Climate Change Rule of Thumb: Cold 'Things' Warming Faster Than Warm Things," National Oceanic and Atmospheric Administration, Novem-

ber 24, 2015, https://www.climate.gov/news-features/blogs/beyond-data/climate
-change-rule-thumb-cold-things-warming-faster-warm-things.

29. N. Pepin et al., "Elevation-Dependent Warming in Mountain Regions of the
World," *Nature Climate Change* 5, no. 5 (May 2015):424–30, https://doi.org/10
.1038/nclimate2563.

30. Arndt, "Climate Change Rule of Thumb."

31. Shaobing Peng et al., "Rice Yields Decline with Higher Night Temperature from
Global Warming," abstract, *Proceedings of the National Academy of Sciences* 101,
no. 27 (July 6, 2004):9971, https://doi.org/10.1073/pnas.0403720101; Ma. Rebecca C. Laza et al., "Differential Response of Rice Plants to High Night Temperatures Imposed at Varying Developmental Phases," abstract, *Agricultural and
Forest Meteorology* 209–210 (September 15, 2015):69, https://doi.org/10.1016/j
.agrformet.2015.04.029.

32. Jerry L. Hatfield, "Increased Temperatures Have Dramatic Effects on Growth and
Grain Yield of Three Maize Hybrids," *Ael* 1, no. 1 (2016):4, https://doi.org/10.2134
/ael2015.10.0006.

33. Tapan B. Pathak and C. Scott Stoddard, "Climate Change Effects on the Processing
Tomato Growing Season in California Using Growing Degree Day Model," abstract, *Modeling Earth Systems and Environment* 4, no. 2 (June 1, 2018):765,
https://doi.org/10.1007/s40808-018-0460-y.

34. "2019 State Agriculture Overview for California," US Department of Agriculture
(USDA) National Agricultural Statistics Service, accessed March 3, 2020, https://
www.nass.usda.gov/Quick_Stats/Ag_Overview/stateOverview.php?state
=CALIFORNIA.

35. Jerry L. Hatfield and John H. Prueger, "Temperature Extremes: Effect on Plant
Growth and Development," *Weather and Climate Extremes* 10(A) (December 1,
2015):4, https://doi.org/10.1016/j.wace.2015.08.001.

36. Delphine Deryng et al., "Global Crop Yield Response to Extreme Heat Stress under
Multiple Climate Change Futures," *Environmental Research Letters* 9, no. 3
(March 2014):9, https://doi.org/10.1088/1748-9326/9/3/034011.

37. Nicola Jones, "Redrawing the Map: How the World's Climate Zones Are Shifting," Yale Environment 360, October 23, 2018, https://e360.yale.edu/features
/redrawing-the-map-how-the-worlds-climate-zones-are-shifting.

38. Eike Luedeling, Minghua Zhang, and Evan H. Girvetz, "Climatic Changes Lead
to Declining Winter Chill for Fruit and Nut Trees in California during 1950–2099,"
PLOS ONE 4, no. 7 (July 22, 2009):1–9, https://doi.org/10.1371/journal.pone
.0006166.

39. Ella Koeze, "How a Warm Winter Destroyed 85 Percent of Georgia's Peaches,"
FiveThirtyEight, September 14, 2017, https://fivethirtyeight.com/features/how-a
-warm-winter-destroyed-85-percent-of-georgias-peaches/.

40. Sarah Zhang, "Time to Add Pistachios to California's List of Woes," *Wired*, September 16, 2015, https://www.wired.com/2015/09/time-add-pistachios-californias
-list-problems/.

41. Dennis Baldocchi and Simon Wong, "Accumulated Winter Chill Is Decreasing in the Fruit Growing Regions of California," abstract, *Climatic Change* 87, no. 1 (March 1, 2008):153, https://doi.org/10.1007/s10584-007-9367-8.

42. Wolfe et al., "Unique Challenges and Opportunities," 236; Alexander G. Peterson and John T. Abatzoglou, "Observed Changes in False Springs over the Contiguous United States," abstract, *Geophysical Research Letters* 41, no. 6 (March 28, 2014):2156, https://doi.org/10.1002/2014GL059266.

43. Franklin D. Roosevelt to all state governors on a Uniform Soil Conservation Law, February 26, 1937, the American Presidency Project, University of California Santa Barbara, https://www.presidency.ucsb.edu/documents/letter-all-state-governors-uniform-soil-conservation-law.

44. "Soil Health," USDA Natural Resource Conservation Service, accessed March 5, 2019, https://www.nrcs.usda.gov/wps/portal/nrcs/main/soils/health/.

45. "Crop Rooting Depth," University of California Division of Agriculture and Natural Resources, accessed April 22, 2019, http://ucmanagedrought.ucdavis.edu/Agriculture/Irrigation_Scheduling/Evapotranspiration_Scheduling_ET/Frequency_of_Irrigation/Crop_Rooting_Depth.

46. E. L. Stone and P. J. Kalisz, "On the Maximum Extent of Tree Roots," *Forest Ecology and Management* 46, no. 1 (December 1, 1991):62, https://doi.org/10.1016/0378-1127(91)90245-Q.

47. Connor R. Fitzpatrick et al., "Assembly and Ecological Function of the Root Microbiome across Angiosperm Plant Species," *Proceedings of the National Academy of Sciences* 115, no. 6 (February 6, 2018):E1157–65, https://doi.org/10.1073/pnas.1717617115.

48. Christina Nunez, "Deforestation Explained," *National Geographic*, February 7, 2019, https://www.nationalgeographic.com/environment/global-warming/deforestation/.

49. Rattan Lal, "Carbon Sequestration," *Philosophical Transactions of the Royal Society B: Biological Sciences* 363, no. 1492 (30 August 2007):816, https://doi.org/10.1098/rstb.2007.2185.

50. C. Le Quéré et al., "Global Carbon Budget 2017," *Earth System Science Data* 10 (March 2018):409, https://doi.org/10.5194/essd-10-405-2018.

51. Sonja Vermeulen et al., "A Global Agenda for Collective Action on Soil Carbon," *Nature Sustainability* 2, no. 1 (January 2019):2, https://doi.org/10.1038/s41893-018-0212-z.

52. "Livestock and Poultry: World Markets and Trade," USDA Foreign Agricultural Service, January 10, 2020, https://apps.fas.usda.gov/psdonline/circulars/livestock_poultry.pdf.

53. Hannah Ritchie, "Half of the World's Habitable Land Is Used for Agriculture," Our World in Data, November 11, 2019, https://ourworldindata.org/global-land-for-agriculture.

54. "Insight into the US Animal Feed Market," AllAboutFeed, January 31, 2018, https://www.allaboutfeed.net/Compound-Feed/Articles/2018/1/Insight-into-the-US-animal-feed-market-242428E/.

55. M. Melissa Rojas-Downing et al., "Climate Change and Livestock: Impacts, Adaptation, and Mitigation," abstract, *Climate Risk Management* 16 (2017):145, https://doi.org/10.1016/j.crm.2017.02.001.

56. Adam J. Vanbergen, "Threats to an Ecosystem Service: Pressures on Pollinators," *Frontiers in Ecology and the Environment* 11, no. 5 (2013):251–59, https://doi.org/10.1890/120126.

57. Simon G. Potts et al., "Global Pollinator Declines: Trends, Impacts and Drivers," abstract, *Trends in Ecology & Evolution* 25, no. 6 (June 2010):345, https://doi.org/10.1016/j.tree.2010.01.007; Gary D. Powney et al., "Widespread Losses of Pollinating Insects in Britain," abstract, *Nature Communications* 10, no. 1 (March 26, 2019):1, https://doi.org/10.1038/s41467-019-08974-9.

58. J. Hatfield et al., "Agriculture," in *Climate Change Impacts in the United States: The Third National Climate Assessment*, ed. J. M. Melillo, Terese (T.C.) Richmond, and G. W. Yohe (Washington, DC: US Global Change Research Program, 2014), 151, https://doi.org/10.7930/J02Z13FR.

59. E.-C. Oerke, "Crop Losses to Pests," *Journal of Agricultural Science* 144, no. 1 (February 2006):31–43, https://doi.org/10.1017/S0021859605005708.

60. James R. Bell et al., "Spatial and Habitat Variation in Aphid, Butterfly, Moth and Bird Phenologies over the Last Half Century," *Global Change Biology* 25, no. 6 (2019):1987, https://doi.org/10.1111/gcb.14592.

61. Philipp Lehmann et al., "Complex Responses of Global Insect Pests to Climate Change," *Front Ecol Environ*, 18, no. 3 (2020):141–50, https://doi.org/10.1101/425488.

62. S. L. Douglass, B. M. Webb, and R. Kilgore, "Highways in the Coastal Environment: Assessing Extreme Events," US Department of Transportation Federal Highway Department, Report FHWA-NHI-14–006, October 2014, 1, https://rosap.ntl.bts.gov/view/dot/41312.

63. Jennifer M. Jacobs et al., "Recent and Future Outlooks for Nuisance Flooding Impacts on Roadways on the U.S. East Coast," abstract, *Transportation Research Record* 2672, no. 2 (December 1, 2018):1, https://doi.org/10.1177/0361198118756366.

64. "Grain Transportation: Flooding Stops MS River Traffic," USDA, March 29, 2019, https://agfax.com/2019/03/29/grain-transportation-flooding-stops-ms-river-traffic/.

65. Rob Bailey and Laura Wellesley, *Chokepoints and Vulnerabilities in Global Food Trade* (London: Chatham House, 2017), iv, https://www.chathamhouse.org/publication/chokepoints-vulnerabilities-global-food-trade.

66. E. Somanathan et al., "The Impact of Temperature on Productivity and Labor Supply: Evidence from Indian Manufacturing," abstract (working paper 2018–69, Becker Friedman Institute, University of Chicago, August 2018), 1, https://epic.uchicago.edu/wp-content/uploads/2019/07/Working-Paper-1.pdf.

67. Deepak K. Ray et al., "Climate Change Has Likely Already Affected Global Food Production," *PLOS ONE* 14, no. 5 (May 31, 2019):1–18, https://doi.org/10.1371/journal.pone.0217148.

68. "About: Agrobiodiversity Linked Data Consortium," Agricultural and Forestry Biodiversity Information Commons, accessed March 9, 2020, http://www.agrobio diversity.org/about.

Beer, Wine, and Spirits

1. Patrick E. McGovern et al., "Fermented Beverages of Pre- and Proto-Historic China," *Proceedings of the National Academy of Sciences* 101, no. 51 (December 21, 2004):17593–98, https://doi.org/10.1073/pnas.0407921102.
2. Brian Hayden, Neil Canuel, and Jennifer Shanse, "What Was Brewing in the Natufian? An Archaeological Assessment of Brewing Technology in the Epipaleolithic," *Journal of Archaeological Method and Theory* 20, no. 1 (March 1, 2013):102–50, https://doi.org/10.1007/s10816-011-9127-y.
3. "Major Barley Producers Worldwide in 2018/19, by Country," Statista, released February 2019, https://www.statista.com/statistics/272760/barley-harvest-forecast/.
4. "Barley Profile," Agricultural Marketing Resource Center, revised October 2018, https://www.agmrc.org/commodities-products/grains-oilseeds/barley-profile.
5. "National Hop Report," US Department of Agriculture (USDA) National Agricultural Statistics Service, 2019, 8, https://www.usahops.org/img/blog_pdf/269.pdf.
6. Kirin Beer University Report Global Beer Consumption by Country in 2017, Kirin Holdings Company, December 20, 2018, https://www.kirinholdings.co.jp/english/news/2018/1220_01.html.
7. Ari Levaux, "Montana: Beer in the Climate Crosshairs," *United States of Climate Change*, Weather Channel, December 13, 2017, https://features.weather.com/us-climate-change/montana/.
8. Wei Xie et al., "Decreases in Global Beer Supply Due to Extreme Drought and Heat," abstract, *Nature Plants* 4, no. 11 (November 2018):964, https://doi.org/10.1038/s41477-018-0263-1.
9. Caitlyn Kennedy, "Climate & Beer," National Oceanic and Atmospheric Administration, January 13, 2016, https://www.climate.gov/news-features/climate-and/climate-beer.
10. Philip W. Mote et al., "Dramatic Declines in Snowpack in the Western US," abstract, *npj Climate and Atmospheric Science* 1, no. 2 (March 2, 2018):1, https://doi.org/10.1038/s41612-018-0012-1.
11. Kennedy, "Climate & Beer."
12. Alastair Bland, "California Brewers Fear Drought Could Leave Bad Taste in Your Beer," National Public Radio, February 20, 2014, https://www.npr.org/sections/thesalt/2014/02/19/279627234/california-brewers-fear-drought-could-leave-bad-taste-in-your-beer.
13. Kennedy, "Climate & Beer."
14. Agence France-Presse, "Climate Change Blamed for Putting Belgium Beer Business at Risk," *Guardian*, November 3, 2015, https://www.theguardian.com/world/2015/nov/04/climate-change-blamed-for-putting-belgium-beer-business-at-risk.

15. "Belgium," Center for Climate Adaptation, updated February 29, 2020, https://www.climatechangepost.com/belgium/climate-change/.

16. Mathias Wiegmann et al., "Barley Yield Formation under Abiotic Stress Depends on the Interplay between Flowering Time Genes and Environmental Cues," *Scientific Reports* 9, no. 1 (April 25, 2019):1–16, https://doi.org/10.1038/s41598-019-42673-1.

17. Levaux, "Beer in the Climate Crosshairs."

18. A. C. Shilton, "Meet the Wild Hop Hunters Saving Your Beer from Climate Change," *Outside*, December 14, 2016, https://www.outsideonline.com/2141991/meet-wild-hop-hunters-saving-your-beer-climate-change.

19. Danil Boparai, "Brewdog Launches New Beer Designed to 'Remind Leaders to Prioritise Climate Change Issues,'" Dezeen, November 11, 2017, https://www.dezeen.com/2017/11/11/make-earth-great-again-brewdog-beer-donald-trump-paris-agreement-climate-change-design/.

20. Kate Sheppard, "Beer Brewers Unite to Call for Action on Climate Change," Huff-Post, March 10, 2015, https://www.huffpost.com/entry/beer-climate-change_n_6839724.

21. Kennedy, "Climate & Beer."

22. Sheppard, "Beer Brewers Unite."

23. Patrick McGovern et al., "Early Neolithic Wine of Georgia in the South Caucasus," abstract, *Proceedings of the National Academy of Sciences* 114, no. 48 (November 28, 2017):E10309, https://doi.org/10.1073/pnas.1714728114.

24. "Wine Production Worldwide from 1990 to 2018," Statista, released April 2019, https://www.statista.com/statistics/397870/global-wine-production/.

25. "Wine Production Worldwide in 2018, by Country," Statista, released April 2019, https://www.statista.com/statistics/240638/wine-production-in-selected-countries-and-regions/.

26. "Per Capita Wine Consumption Worldwide in 2014 and 2018, by Country," Statista, released February 2015, https://www.statista.com/statistics/232754/leading-20-countries-of-wine-consumption/.

27. "Wine Industry Economic Impact Reports," National Association of American Wineries, 2017, https://wineamerica.org/impact/.

28. Wine Communication Group, "January 2019: Total U.S. Wine Market," Wine Analytics Report, January 2019, https://wineanalyticsreport.com/report/january-2019-total-u-s-wine-market/; Wine Communication Group, "Wines Vines Analytics," Wine Analytics Report, accessed August 27, 2019, https://winesvinesanalytics.com/statistics/winery/.

29. Ramón Mira de Orduña, "Climate Change Associated Effects on Grape and Wine Quality and Production," *Food Research International* 43, no. 7 (August 1, 2010):1851, https://doi.org/10.1016/j.foodres.2010.05.001.

30. Benjamin I. Cook and Elizabeth M. Wolkovich, "Climate Change Decouples Drought from Early Wine Grape Harvests in France," *Nature Climate Change* 6, no. 7 (2016):716, https://www.nature.com/articles/nclimate2960.

31. "Extreme Weather Events in Europe," European Academies Science Advisory Council, March 21, 2018, 2, https://issuu.com/easaceurope/docs/easac_statement _extreme_weather_eve.

32. "Global Wine Output 'to Hit 50-Year Low,'" BBC News, October 24, 2017, https:// www.bbc.com/news/business-41728536.

33. Andrew Jefford, "'2017 Must Be One of the Most Disaster-Strewn Years since *phyl-loxera*,'" *Decanter*, February 16, 2018, https://www.decanter.com/premium/andrew -jefford-2017-vintage-383485/.

34. Esther Mobley, "Hope in Wine Country as Vineyards Assess the Long-Term Economic Impact of the Wildfires," SFGate, October 14, 2017, http://www.sfgate.com /wine/article/Hope-in-Wine-Country-as-vineyards-assess-the-12278600.php.

35. John Kell, "How the Wine Industry Is Reckoning with the Impending Effects of California Wildfires," *Fortune*, November 6, 2019, https://fortune.com/2019/11/06 /california-wildfires-wine-industry-kincade-fire-sonoma/.

36. L. Hannah et al., "Climate Change, Wine, and Conservation," abstract, *Proceedings of the National Academy of Sciences* 110, no. 17 (April 23, 2013):6907, https:// doi.org/10.1073/pnas.1210127110.

37. M. A. White et al., "Extreme Heat Reduces and Shifts United States Premium Wine Production in the 21st Century," abstract, *Proceedings of the National Academy of Sciences* 103, no. 30 (July 25, 2006):11217, https://doi.org/10.1073/pnas.0603230103.

38. Gregory V. Jones et al., "Climate Change and Global Wine Quality," abstract, *Climatic Change* 73, no. 3 (December 2005):319, https://doi.org/10.1007/s10584-005 -4704-2; Cornelis van Leeuwen and Philippe Darriet, "The Impact of Climate Change on Viticulture and Wine Quality," *Journal of Wine Economics* 11, no. 1 (May 2016):164, https://doi.org/10.1017/jwe.2015.21.

39. Dennis Baldocchi and Simon Wong, "Accumulated Winter Chill Is Decreasing in the Fruit Growing Regions of California," *Climatic Change* 87, no. 1 (March 1, 2008):155, https://doi.org/10.1007/s10584-007-9367-8.

40. "Grape: Pierce's Disease," University of California Statewide Integrated Pest Management Program, accessed August 15, 2019, http://ipm.ucanr.edu/PMG/r302101211 .html.

41. A. B. Tate, "Global Warming's Impact on Wine," *Journal of Wine Research* 12, no. 2 (August 2001):100, https://doi.org/10.1080/09571260120095012.

42. A. Costa et al., "Climate Response of Cork Growth in the Mediterranean Oak (*Quercus suber* L.) Woodlands of Southwestern Portugal," abstract, *Dendrochronologia* 38, supplement C (June 1, 2016):72, https://doi.org/10.1016/j.dendro.2016.03.007.

43. Alex Whiting, "Europe's Wine Industry May Suffer with Global Warming—Research," Reuters, July 13, 2017, https://www.reuters.com/article/us-climatechange -wine-europe-idUSKBN19Y24D.

44. "Fresh Apples, Grapes, and Pears: World Markets and Trade," USDA Foreign Agricultural Service, December 2019, 4, https://apps.fas.usda.gov/psdonline/circulars /fruit.pdf.

45. Hannah et al., "Climate Change, Wine, and Conservation," 6908.

46. David W. Wolfe et al., "Unique Challenges and Opportunities for Northeastern US Crop Production in a Changing Climate," *Climatic Change* 146, no. 1 (January 1, 2018):239, https://doi.org/10.1007/s10584-017-2109-7.

47. Gregory V. Jones and Hans R. Schultz, "Climate Change and Emerging Cool Climate Wine Regions," *Wine and Viticulture Journal* (November/December 2016):51–53, https://www.linfield.edu/assets/files/Wine-Studies/GregJones/Jones__Schultz_NovDec2016WVJ.pdf.

48. Simon Royal, "South Australia's Famed Wine Regions Preparing for the Squeeze of Climate Change," ABC News, April 14, 2018, https://www.abc.net.au/news/2018-04-15/australias-wine-regions-feeling-pinch-of-climate-change/9644660.

49. E. M. Wolkovich et al., "From Pinot to Xinomavro in the World's Future Wine-Growing Regions," abstract, *Nature Climate Change* 8, no. 1 (January 2018):29, https://doi.org/10.1038/s41558-017-0016-6.

50. Peter Reuell, "As Climate Changes, So Will Wine Grapes," *Harvard Gazette*, January 9, 2018, https://news.harvard.edu/gazette/story/2018/01/as-climate-changes-so-does-wine/.

51. California Sustainable Winegrowing Alliance, "A Winegrowers' Guide to Navigating Risks," n.d., https://www.sustainablewinegrowing.org/docs/Risk_Guide_Second_Edition.pdf.

52. David Gelles, "Falcons, Drones, Data: A Winery Battles Climate Change," *New York Times*, January 5, 2017, https://www.nytimes.com/2017/01/05/business/california-wine-climate-change.html.

53. Torres Wines, "Torres and Earth," April 5, 2017, https://www.torres.es/en/blog/wine-planet/torres-earth.

54. Wine Industry Network, "International Wineries for Climate Action Named Social Visionary of the Year by Wine Enthusiast," Wine Industry Network Advisor, November 6, 2019, https://wineindustryadvisor.com/2019/11/06/international-wineries-climate-action-named-social-visionary.

55. Sandra Taylor, "Driven by Consumers, Climate Change, Sustainability Grows in the Global Wine Industry," SmartBrief, January 3, 2018, https://www.smartbrief.com/original/2018/01/driven-consumers-climate-change-sustainability-grows-global-wine-industry.

56. "Spirits: Worldwide," Statista, released August 2019, https://www.statista.com/outlook/10020000/100/spirits/worldwide.

57. "Distilled Spirits Consumption Worldwide in 2015, by Leading Countries," Statista, released July 2017, https://www.statista.com/statistics/270218/spirits-consumption-worldwide-by-country/.

58. "Sales Market Share of the United States Alcohol Industry from 2000 to 2019, by Beverage," Statista, released February 2019, https://www.statista.com/statistics/233699/market-share-revenue-of-the-us-alcohol-industry-by-beverage/.

59. "Per Capita Consumption of Distilled Spirits in the United States in 2018, by State," Statista, released July 2019, https://www.statista.com/statistics/224581/per -capita-spirit-consumption-in-the-us-in-2010-by-state/.

60. "Dollar Sales of Distilled Spirits in the United States in 2018, by Tier," Statista, released July 2019, https://www.statista.com/statistics/462820/us-dollar-sales-of -distilled-spirits-by-tier/.

61. "Sales Volume of the United States Spirits Industry from 2010 to 2019, by Category," Statista, released February 2019, https://www.statista.com/statistics/233527 /sales-volume-of-the-us-spirits-industry-by-category/.

62. Tom Kimmerer, "Bourbon, Barrels and Climate," Planet Experts, December 30, 2014, http://www.planetexperts.com/bourbon-barrels-climate/.

63. Lindsay Brandon, "Hot Scotch: The Impact of Climate Change on Your Whisky," Whiskey Wash, June 26, 2017, https://thewhiskeywash.com/lifestyle/hot-scotch -impact-climate-change-whisky/.

64. Simon Roach, "Scotch on the Rocks: Distilleries Fear Climate Crisis Will Endanger Whisky Production," *Guardian*, June 2, 2019, https://www.theguardian.com /uk-news/2019/jun/02/scotland-whisky-climate-crisis-heatwave-distilleries-halt -production.

65. Nick Hines, "The Amount of Scotch Lost to the Angel's Share Every Year Is Staggering," *VinePair* (blog), April 11, 2017, https://vinepair.com/articles/what-is-angels -share-scotch/.

66. "Sales Volume of Tequila in the United States from 2004 to 2019," Statista, released February 2019, https://www.statista.com/statistics/311633/us-sales-volume-of -tequila/.

67. "Tequila Production in Mexico from 1995 to 2019," Statista, released January 2019, https://www.statista.com/statistics/311696/mexico-s-tequila-production/; "Leading Countries of Destination of Tequila Exports from Mexico in 2019," Statista, released February 2020, https://www.statista.com/statistics/311749/mexico-s-export -quantity-of-tequila-by-country/.

68. Remigio Madrigal-Lugo, "Mexican Scientists Adapt Agave Production in Response to Climate Change," El Tecolote, May 17, 2018, http://eltecolote.org/content /en/features/mexican-scientists-adapt-agave-production-in-response-to-climate -change/.

69. Anthony D. Blue, *The Complete Book of Spirits: A Guide to Their History, Production, and Enjoyment* (New York: William Morrow, 2004):9.

70. Juhwan Lee, Steven De Gryze, and Johan Six, "Effect of Climate Change on Field Crop Production in California's Central Valley," abstract, *Climatic Change* 109 (December 2011):335, https:// doi 10.1007/s10584-011-0305-4.

71. David M. Checkley, Rebecca G. Asch, and Ryan R. Rykaczewski, "Climate, Anchovy, and Sardine," abstract, *Annual Review of Marine Science* 9, no. 1 (2017):469, https://doi.org/10.1146/annurev-marine-122414-033819.

72. M. Alizany et al., "Adapting to Cyclones in Madagascar's Analanjirofo Region," *Adaptation Insights* 7 (November 2010):1–4, https://assets.publishing.service.gov

.uk/media/57a08b0bed915d3cfd000ace/Adaptation-Insight-Madagascar
-Adapting-to-cyclones.pdf.

73. Oliver Milman, "Hotting up: How Climate Change Could Swallow Louisiana's Tabasco Island," *Guardian*, March 27, 2018, https://www.theguardian.com/envi ronment/2018/mar/27/climate-change-louisiana-tabasco-avery-island.

74. Stephen Daniells, "The Erosion of Predictability: Climate Change and the Botani cal Supply Chain," updated April 17, 2019, https://www.nutraingredients-usa.com /Article/2018/03/22/The-erosion-of-predictability-Climate-change-and-the-botani cal-supply-chain.

75. Amjad M. Husaini, "Challenges of Climate Change," abstract, *GM Crops & Food* 5, no. 2 (April 11, 2014):97, https://doi.org/10.4161/gmcr.29436.

76. Suddhasuchi Das and Amit Baran Sharangi, "Impact of Climate Change on Spice Crops," in *Indian Spices: The Legacy, Production and Processing of India's Trea sured Export*, ed. Amit Baran Sharangi (Cham, Switzerland: Springer, 2018), 379– 404, https://doi.org/10.1007/978-3-319-75016-3_14.

77. Science Daily, "How Tequila Could Be Key in Our Battle against Climate Change," *Science Daily*, December 6, 2016, https://www.sciencedaily.com/releases/2016 /12/161206094243.htm.

78. Rachel King. "How Wineries and Distilleries Are Addressing Climate Change," *Fortune*, October 1, 2018, https://fortune.com/2018/10/01/wineries-distilleries -climate-change/.

79. Danielle MacDonald, "Sustainable Spirits," Alcohol Professor, accessed Au gust 12, 2019, https://www.alcoholprofessor.com/blog-posts/blog/2018/06/16/sus tainable-spirits.

80. Connie Baker, as told to Francine Mourakian, "This Zero-Waste Distillery Saves 4 Million Gallons of Water per Year—and Still Makes Delicious Stuff," *Popular Mechanics*, June 9, 2019, https://www.popularmechanics.com/home/food-drink /a27006051/marble-distilling-sustainable-spirits/.

Salads

1. "Salad," Encyclopedia.com, updated March 2, 2020, https://www.encyclopedia .com/sports-and-everyday-life/food-and-drink/food-and-cooking/salads.

2. "Salad," Encyclopedia.com.

3. Hoang Nguyen, "Most Americans Are Fans of Salads," YouGov, March 16, 2018, https://today.yougov.com/topics/food/articles-reports/2018/03/19/most -americans-are-fans-salads.

4. David W. Wolfe et al., "Unique Challenges and Opportunities for Northeastern US Crop Production in a Changing Climate," *Climatic Change* 146, no. 1 (Janu ary 1, 2018):232, https://doi.org/10.1007/s10584-017-2109-7.

5. David W. Wolfe et al., "Projected Change in Climate Thresholds in the Northeast ern U.S.: Implications for Crops, Pests, Livestock, and Farmers," abstract, *Mitigation*

and *Adaptation Strategies for Global Change* 13, no. 5 (June 1, 2008):555, https://doi.org/10.1007/s11027-007-9125-2.

6. Jinlong Dong et al., "Effects of Elevated CO_2 on Nutritional Quality of Vegetables: A Review," *Frontiers in Plant Science* 9 (2018):1–11, https://doi.org/10.3389/fpls.2018.00924.

7. Komivi Dossa et al., "Enhancing Sesame Production in West Africa's Sahel: A Comprehensive Insight into the Cultivation of This Untapped Crop in Senegal and Mali," abstract, *Agriculture & Food Security* 6, no. 1 (December 14, 2017):1, https://doi.org/10.1186/s40066-017-0143-3.

8. "Per Capita Consumption of Fresh Avocados in the United States from 2000 to 2018," Statista, released October 2019, https://www.statista.com/statistics/257192/per-capita-consumption-of-fresh-avocados-in-the-us/.

9. "Domestic Avocado Consumption in the United States from 1985 to 2019," Statista, released October 2019, https://www.statista.com/statistics/591263/average-avocado-consumption-us-per-week/; "Super Bowl Avocados Consumption," Southeast AgNet Radio, January 29, 2019, http://southeastagnet.com/2019/01/29/super-bowl-avocados-consumption/.

10. "Consumers' Main Drivers for Buying Avocados in the United States in 2019," Statista, modified September 2019, https://www.statista.com/statistics/317753/us-consumers--main-drivers-for-buying-avocados/.

11. Megan Ware, "Why Is Avocado Good for You?," Medical News Today, accessed November 21, 2019, https://www.medicalnewstoday.com/articles/270406.php; "Potassium Content of Foods List," Drugs.com, updated September 24 2019, https://www.drugs.com/cg/potassium-content-of-foods-list.html.

12. Lawrence Marais, "Avocado Diseases of Major Importance Worldwide and Their Management," *Diseases of Fruits and Vegetables*, 2 (2007):1, https://doi.org/10.1007/1-4020-2607-2_1.

13. "Crops," Food and Agriculture Organization of the United Nations (FAO), 2017, http://www.fao.org/faostat/en/#data/QC.

14. "Global Production of Avocados in 2018, by Country," Statista, released 2020, https://www.statista.com/statistics/593211/global-avocado-production-by-country/.

15. "Category Share of Avocado Sales in the United States in 2019, by Avocado Type," Statista, released September 2019, https://www.statista.com/statistics/191355/fresh-avocado-category-share-in-2011/.

16. "Production of Avocados in the United States in 2018, by State," Statista, released May, 2020, https://www.statista.com/statistics/610460/production-avocados-us-by-state/.

17. "Avocado," Tridge, accessed March 3, 2020, tridge.com/intelligences/avocado/import; "Avocados," Agricultural Marketing Resource Center, revised October 2018, https://www.agmrc.org/commodities-products/fruits/avocados.

18. Gary Bender, ed., *Avocado Production in California: A Cultural Handbook for Growers*, *Book One* (University of California Cooperative Extension and Califor-

nia Avocado Society, n.d.), 45, http://ucanr.edu/sites/alternativefruits/files/221200
.pdf.

19. Bender, *Avocado Production*, 39.

20. Transparency Market Research, "Avocado Market to Bag $21.56 Billion with Rising Preference for Healthy Food," PR Newswire, September 20, 2018, https://www
.prnewswire.com/news-releases/avocado-market-to-bag-us-21-56-billion-with-rising-preference-for-healthy-food-among-vastly-expanding-population-tmr-874
387070.html.

21. "Global Fruit Production in 2018 by Selected Variety," Statista, released February 2020, https://www.statista.com/statistics/264001/worldwide-production-of
-fruit-by-variety/.

22. Charlotte Mackenzie, June 22, 2016, "Latin America, Feeding the Global Avocado
Obsession," *Farm Folio Blog*, https://farmfolio.net/articles/latin-america-feeding
-global-avocado-obsession/.

23. Tapan Pathak et al., "Climate Change Trends and Impacts on California Agriculture: A Detailed Review," *Agronomy* 8, no. 3 (February 26, 2018):11, https://doi.org
/10.3390/agronomy8030025.

24. Bender, *Avocado Production*, 57.

25. Nicholas Casey, "In Peru's Deserts, Melting Glaciers Are a Godsend (Until They're
Gone)," *New York Times*, November 26, 2017, https://www.nytimes.com/2017
/11/26/world/americas/peru-climate-change.html.

26. M. Carey, "Glacier Runoff and Human Vulnerability to Climate Change: The Case
of Export Agriculture in Peru," American Geophysical Union, abstract, GC21E-02
(December 1, 2013), http://adsabs.harvard.edu/abs/2013AGUFMGC21E..02C.

27. Alice Facchini and Sandra Laville, "Chilean Villagers Claim British Appetite for
Avocados Is Draining Region Dry," *Guardian*, May 17, 2018, https://www.the
guardian.com/environment/2018/may/17/chilean-villagers-claim-british-appetite
-for-avocados-is-draining-region-dry.

28. Eric M. Bell, Jennifer R. Stokes-Draut, and Arpad Horvath, "Environmental Evaluation of High-Value Agricultural Produce with Diverse Water Sources: Case
Study from Southern California," abstract, *Environmental Research Letters* 13
(2018):1, https://doi.org/10.1088/1748-9326/aaa49a.

29. David B. Lobell, Christopher B. Field, and Kimberly Nicholas Cahill, "Historical
Effects of Temperature and Precipitation on California Crop Yields," *Climatic
Change* 81, no. 2 (March 2007):198, https://doi.org/10.1007/s10584-006-9141-3.

30. David B. Lobell et al., "Impacts of Future Climate Change on California Perennial Crop Yields: Model Projections with Climate and Crop Uncertainties," *Agricultural and Forest Meteorology* 141, no. 2 (December 20, 2006):208, 215, https://doi
.org/10.1016/j.agrformet.2006.10.006.

31. Xiomara N. D. Caballero and Gabriela G. B. Flores, "Forests Falling Fast to Make
Way for Mexican Avocado," Global Forest Watch, March 20, 2019, https://blog
.globalforestwatch.org/commodities/forests-falling-fast-to-make-way
-for-mexican-avocado.

32. "In Mexico, It's Avocado Farms vs. the Forest," CBS News, August 10, 2016, https://www.cbsnews.com/news/in-mexico-its-avocado-farms-vs-the-forest/.

33. Ashley Nickle, "Effects of Hurricane Linger for Florida Avocados," Packer, May 27, 2018, https://www.thepacker.com/article/effects-hurricane-linger-florida-avocados.

34. "Diseases of Avocado," American Phytopathological Society, updated April 8, 2003, https://www.apsnet.org/edcenter/resources/commonnames/Pages/Avocado.aspx.

35. University of California Statewide Integrated Pest Management Program, "Phytophthora Root Rot," University of California Agriculture and Natural Resources, last modified September 2016, http://ipm.ucanr.edu/PMG/r8100111.html; Marais, "Avocado Diseases of Major Importance," 2.

36. Bender, "Avocado Production," 48.

37. M. Montserrat et al., "Pollen Supply Promotes, but High Temperatures Demote, Predatory Mite Abundance in Avocado Orchards," abstract, *Agriculture, Ecosystems & Environment* 164 (2013):155, https://doi.org/10.1016/j.agee.2012.09.014.

38. Cesar Marchioro, "Global Potential Distribution of *Bactrocera carambolae* and the Risks for Fruit Production in Brazil," *PLOS ONE* 11 (2016):9, https://doi.org/10.1371/journal.pone.0166142.

39. "REDD+ Reducing Emissions from Deforestation and Forest Degradation," FAO, accessed March 4, 2020, http://www.fao.org/redd/en/.

40. Deanna Newsom and Jeffrey C. Milder, *2018 Rainforest Alliance Impacts Report: Partnership, Learning, and Change* (Rainforest Alliance, 2018), 7, https://www.rainforest-alliance.org/sites/default/files/2018-03/RA_Impacts_2018.pdf.

41. Deanroy Mbabazi et al., "An Irrigation Schedule Testing Model for Optimization of the Smartirrigation Avocado App," *Agricultural Water Management* 179 (January 1, 2017):396, https://doi.org/10.1016/j.agwat.2016.09.006.

42. Jane Muller et al., *Climate Change and Climate Policy Implications for the Australian Avocado Industry* (Sydney: Horticulture Australia, 2010), 51.

43. Lesley McClurg, "New Growing Technique Relieves Drought Stricken Avocado Farmers," Capital Radio, June 3, 2015, http://www.capradio.org/50437; Jonathan Crane, Tropical Research and Education Center, University of Florida, April 1, 2019, and Ben Faber, Cooperative Extension, University of California, March 9, 2019, personal communication.

44. Jonathan Crane, April 1, 2019, personal communication.

45. Paul Vossen, "Olive Oil: History, Production, and Characteristics of the World's Classic Oils," abstract, *HortScience* 42, no. 5 (August 1, 2007):1093, https://doi.org/10.21273/HORTSCI.42.5.1093.

46. Vossen, "Olive Oil," 1095.

47. European Commission, "Olives by Production," Eurostat, 2018, http://ec.europa.eu/eurostat/tgm/refreshTableAction.do?tab=table&plugin=1&pcode=tag00122&language=en.

48. P. Lazicki and D. Geissler, "Olive Production in California," Univ. of Calif., Davis (2016), https://apps1.cdfa.ca.gov/FertilizerResearch/docs/Olive_Production_CA.pdf;

"Olive Oil Production by Country," North American Olive Oil Association, April 21, 2015, https://www.aboutoliveoil.org/olive-oil-production-by-country.

49. Vossen, "Olive Oil," 1095.

50. Dan Flynn, University of California, Davis, January 25, 2019 personal communication; Rebecca Rupp, "The Bitter Truth About Olives," *National Geographic*, July 1, 2016, https://www.nationalgeographic.com/people-and-culture/food/the-plate/2016/07/olives--the-bitter-truth/; "Table Olives vs. Olive Oil Olives," California Olive Ranch, September 9, 2015, https://californiaoliveranch.com/table-olives-vs-olive-oil-olives/.

51. "Extra Virgin Olive Oil," Olive Oil Times, accessed November 23, 2019, https://www.oliveoiltimes.com/extra-virgin-olive-oil; "Olive Oils," Olive Oil Source, 2019, https://www.oliveoilsource.com/page/product-grade-definitions.

52. Saad Fayed, "Understanding the Different Types of Olive Oil," The Spruce, Eats, updated November 18, 2018, https://www.thespruceeats.com/olive-oil-2355732.

53. "U.S. Olive Oil Consumption in the United States from 2000 to 2019," Statista, released January 2020, https://www.statista.com/statistics/288368/olive-oil-consumption-united-states/.

54. "Olive Oil Market Value Worldwide in 2018 and 2028," Statista, released January 2019, https://www.statista.com/statistics/977988/global-olive-oil-market-value/.

55. "Do You Consider Olive Oil to Be Healthy?," Statista, released July 2016. http://www.statista.com/statistics/591904/olive-oil-considered-healthy-by-us-consumers/.

56. Genevieve Buckland and Carlos A. Gonzalez, "The Role of Olive Oil in Disease Prevention: A Focus on the Recent Epidemiological Evidence from Cohort Studies and Dietary Intervention Trials," abstract, *British Journal of Nutrition* 113, no. S2 (April 2015):S94, https://doi.org/10.1017/S0007114514003936.

57. Lina Trabelsi et al., "Impact of Drought and Salinity on Olive Water Status and Physiological Performance in an Arid Climate," abstract, *Agricultural Water Management* 213 (March 1, 2019):749, https://doi.org/10.1016/j.agwat.2018.11.025.

58. Rafaela Dios-Palomares and José M. Martínez-Paz, "Technical, Quality and Environmental Efficiency of the Olive Oil Industry," abstract, *Food Policy* 36, no. 4 (August 1, 2011):526, https://doi.org/10.1016/j.foodpol.2011.04.001.

59. Lazar Tanasijevic et al., "Impacts of Climate Change on Olive Crop Evapotranspiration and Irrigation Requirements in the Mediterranean Region," abstract, *Agricultural Water Management* 144 (October 1, 2014):54, https://doi.org/10.1016/j.agwat.2014.05.019.

60. Vossen, "Olive Oil," 1094.

61. "Turkey Braces for Economic Impact of Erratic Weather," Olive Oil Times, September 9, 2014, https://www.oliveoiltimes.com/production/turkey-braces-economic-impact-erratic-weather/41126.

62. Arthur Neslen, "Italy Sees 57% Drop in Olive Harvest as Result of Climate Change, Scientist Says," *Guardian*, March 5, 2019, https://www.theguardian.com/world/2019/mar/05/italy-may-depend-on-olive-imports-from-april-scientist-says.

63. Inmaculada Romero et al., "Study of Volatile Compounds of Virgin Olive Oils with 'Frostbitten Olives' Sensory Defect," abstract, *Journal of Agricultural and Food Chemistry* 65, no. 21 (May 31, 2017):4314, https://doi.org/10.1021/acs.jafc.7b00712.

64. Neslen, "Italy Sees 57% Drop."

65. Jessica Wapner, "Half of World's Olive Oil Threatened by Deadly Bacteria in Spain," *Newsweek*, July 6, 2017, https://www.newsweek.com/half-worlds-olive -oil-threatened-deadly-bacteria-spain-633020.

66. Dennis Baldocchi and Simon Wong, "Accumulated Winter Chill Is Decreasing in the Fruit Growing Regions of California," abstract, *Climatic Change* 87, no. 1 (March 1, 2008):153, https://doi.org/10.1007/s10584-007-9367-8.

67. Helder Fraga, Joaquim G. Pinto, and João A. Santos, "Climate Change Projections for Chilling and Heat Forcing Conditions in European Vineyards and Olive Orchards: A Multi-model Assessment," abstract, *Climatic Change* 152, no. 1 (January 2019):179, https://doi.org/10.1007/s10584-018-2337-5.

68. Daniel Dawson, "Unusual Weather Leads to Dismal Harvest in California," Olive Oil Times, October 2, 2018, https://www.oliveoiltimes.com/olive-oil-business /unusual-weather-leads-to-dismal-harvest-in-california/64600.

69. Luigi Ponti et al., "Fine-Scale Ecological and Economic Assessment of Climate Change on Olive in the Mediterranean Basin Reveals Winners and Losers," abstract, *Proceedings of the National Academy of Sciences* 111, no. 15 (April 15, 2014):5598, https://doi.org/10.1073/pnas.1314437111.

70. Somini Sengupta, "How Climate Change is Playing Havoc with Olive Oil (and Farmers)," *New York Times*, October 24, 2017, https://www.nytimes.com/2017 /10/24/climate/olive-oil.html.

71. Frank Zalom et al., "Olive Fruit Fly," University of California Statewide Integrated Pest Management Program (2017), http://ipm.ucanr.edu/PMG/PESTNOTES /pn74112.html.

72. Andrew P. Gutierrez, Luigi Ponti, and Q. Cossu, "Effects of Climate Warming on Olive and Olive Fly (*Bactrocera oleae* (Gmelin)) in California and Italy," abstract, *Climatic Change* 95, no. 1 (July 1, 2009):195. https://doi.org/10.1007/s10584-008-9528-4.

73. José E. Fernández et al., "A Regulated Deficit Irrigation Strategy for Hedgerow Olive Orchards with High Plant Density," *Plant and Soil* 372, no. 1 (November 1, 2013):282, https://doi.org/10.1007/s11104-013-1704-2.

74. Lorite et al., "Evaluation of Olive Response and Adaptation Strategies to Climate Change under Semi-arid Conditions," abstract, *Agricultural Water Management* 204 (May 31, 2018):247, https://doi.org/10.1016/j.agwat.2018.04.008.

75. "Intercropping of Olive Groves in Greece," Agforward, n.d., https://www.agforward .eu/index.php/en/intercropping-of-olive-groves-in-greece.html.

76. Tanasijevic et al., "Impacts of Climate Change," 54.

77. "Oregon Sees Potential in a New Crop: Olives," Western Farmer-Stockman, May 9, 2017, https://www.westernfarmerstockman.com/crops/oregon-sees-potential-new -crop-olives.

78. Sengupta, "How Climate Change Is Playing Havoc."

The Main Course

1. Briana Pobiner, "Meat-Eating among the Earliest Humans," *American Scientist* 104, no. 2 (2016):110, https://doi.org/10.1511/2016.119.110.

2. Laura Wyness, "The Role of Red Meat in the Diet: Nutrition and Health Benefits," *Proceedings of the Nutrition Society* 75, no. 3 (August 2016):227–29, https://doi.org/10.1017/S0029665115004267.

3. *World Livestock: Transforming the Livestock Sector through the Sustainable Development Goals* (Rome: Food and Agriculture Organization of the United Nations [FAO], 2018), viii, http://www.fao.org/3/ca1201en/CA1201EN.pdf.

4. "Sector at a Glance," US Department of Agriculture (USDA) Economic Research Service, updated August 28, 2019, https://www.ers.usda.gov/topics/animal-products/cattle-beef/sector-at-a-glance/.

5. "Per Capita Consumption of Beef in the United States from 2018 to 2028," Statista, released February 2019, https://www.statista.com/statistics/183539/per-capita-consumption-of-beef-in-the-us/.

6. "Livestock and Poultry: World Markets and Trade," USDA Foreign Agricultural Service, January 10, 2020, https://apps.fas.usda.gov/psdonline/app/index.html#/app/downloads.

7. James S. Drouillard, "Current Situation and Future Trends for Beef Production in the United States of America—a Review," *Asian-Australasian Journal of Animal Sciences* 31, no. 7 (July 2018):1008, https://doi.org/10.5713/ajas.18.0428.

8. "Greenhouse Gas Emissions per 100 Grams of Protein," Our World in Data, accessed June 29, 2020, https://ourworldindata.org/grapher/ghg-per-protein-poore.

9. "Land Use per Gram of Protein, by Food Type," Our World in Data, accessed March 27, 2019, https://ourworldindata.org/grapher/exports/land-use-per-gram-of-protein-by-food-type.svg.

10. "Industry Statistics," National Cattlemen's Beef Association, accessed August 25, 2019, https://www.ncba.org/beefindustrystatistics.aspx.

11. "Sector at a Glance," USDA Economic Research Service.

12. James S. Drouillard, "Current Situation and Future Trends for Beef Production in the United States of America—a Review," *Asian-Australasian Journal of Animal Sciences* 31, no. 7 (June 21, 2018):1008, https://doi.org/10.5713/ajas.18.0428.

13. "Corn Production by State 2020," World Population Review, February 17, 2020, http://worldpopulationreview.com/states/corn-production-by-state/.

14. *Back to Grass: The Market Potential for U.S. Grassfed Beef* (Stone Barns Center for Food & Agriculture, Bonterra Partners, and SLM, April 2017), 5–6, https://www.stonebarnscenter.org/wp-content/uploads/2017/10/Grassfed_Full_v2.pdf.

15. Matthew N. Hayek and Rachael D. Garrett, "Nationwide Shift to Grass-Fed Beef Requires Larger Cattle Population," *Environmental Research Letters* 13, no. 8 (July 2018):084005, 4, 6, https://doi.org/10.1088/1748-9326/aad401.

16. Paige L. Stanley et al., "Impacts of Soil Carbon Sequestration on Life Cycle Greenhouse Gas Emissions in Midwestern USA Beef Finishing Systems," abstract, *Agricultural Systems* 162 (May 1, 2018):249, https://doi.org/10.1016/j.agsy.2018.02.003.

17. "Managed Grazing," Project Drawdown, accessed August 26, 2019, https://www.drawdown.org/solutions/managed-grazing.

18. Tara Garnett et al., "Grazed and Confused?," Food Climate Research Network, accessed March 5, 2020, 4, https://fcrn.org.uk/sites/default/files/project-files/fcrn_gnc_summary.pdf.

19. Gregory P. Asner et al., "Grazing Systems, Ecosystem Responses, and Global Change," abstract, *Annual Review of Environment and Resources* 29, no. 1 (October 21, 2004):261, https://doi.org/10.1146/annurev.energy.29.062403.102142.

20. "Silvopasture," Association for Temperate Agroforestry, accessed August 28, 2019, https://www.aftaweb.org/about/what-is-agroforestry/silvopasture.html.

21. "Silvopasture," Project Drawdown, accessed May 15, 2019, https://drawdown.org/solutions/silvopasture.

22. M. Melissa Rojas-Downing et al., "Climate Change and Livestock: Impacts, Adaptation, and Mitigation," *Climate Risk Management* 16 (2017):145–63, https://doi.org/10.1016/j.crm.2017.02.001.

23. N. R. St-Pierre, B. Cobanov, and G. Schnitkey, "Economic Losses from Heat Stress by US Livestock Industries," abstract, *Journal of Dairy Science* 86, supplement (June 1, 2003):E52, https://doi.org/10.3168/jds.S0022-0302(03)74040-5.

24. "Overview of the United States Cattle Industry," USDA National Agricultural Statistics Service, June 24, 2016, 8, https://downloads.usda.library.cornell.edu/usda-esmis/files/8s45q879d/9z903258h/qz20swo9p/USCatSup-06-24-2016.pdf.

25. Stephanie Strom, "A Stubborn Drought Tests Texas Ranchers," *New York Times*, April 5, 2013, https://www.nytimes.com/2013/04/06/business/a-long-drought-tests-texas-cattle-ranchers-patience-and-creativity.html.

26. James S. Drouillard, "Current Situation and Future Trends," 1010.

27. Raluca Mateescu, "Genomics of Thermotolerance—Helping Beef Cattle Adapt to the Climate Change," *Open Access Government* (blog), July 16, 2018, https://www.openaccessgovernment.org/genomics-of-thermotolerance-helping-beef-cattle-adapt-to-the-climate-change/36441/.

28. T. L. Mader, "Environmental Stress in Confined Beef Cattle," *Journal of Animal Science* 81, no. 14, supplement 2 (February 1, 2003):E112, https://academic.oup.com/jas/article-abstract/81/14_suppl_2/E110/4789865?redirectedFrom=PDF.

29. Zhengxia Dou, John D. Toth, and Michael L. Westendorf, "Food Waste for Livestock Feeding: Feasibility, Safety, and Sustainability Implications," abstract, *Global Food Security* 17 (June 1, 2018):154, https://doi.org/10.1016/j.gfs.2017.12.003.

30. R. John Wallace et al., "A Heritable Subset of the Core Rumen Microbiome Dictates Dairy Cow Productivity and Emissions," abstract, *Science Advances* 5, no. 7 (July 1, 2019):1, https://doi.org/10.1126/sciadv.aav8391.

31. Breanna M. Roque et al., "Inclusion of *Asparagopsis armata* in Lactating Dairy Cows' Diet Reduces Enteric Methane Emission by over 50 Percent," abstract, *Jour-*

nal of Cleaner Production 234 (October 10, 2019):132, https://doi.org/10.1016/j .jclepro.2019.06.193.

32. S. L. Woodward, G. C. Waghorn, and P. G. Laboyrie, "Condensed Tannins in Birds-foot Trefoil (*Lotus corniculatus*) Reduce Methane Emissions from Dairy Cows," abstract, *Proceedings of the New Zealand Society of Animal Production* 64 (2004):160, http://www.nzsap.org/system/files/proceedings/ab04039.pdf.

33. "A Chicken for Every Pot," political ad, *New York Times*, October 30, 1928, National Archives and Records Administration, https://catalog.archives.gov/id /187095.

34. Dale Wiehoff, "How the Chicken of Tomorrow Became the Chicken of the World," Institute for Agriculture & Trade Policy, March 26, 2013, https://www .iatp.org/blog/201303/how-the-chicken-of-tomorrow-became-the-chicken-of -the-world.

35. "Per Capita Meat Consumption Worldwide in 2015 and 2030, by Type," Statista, released December 2015, https://www.statista.com/statistics/502294/global-meat -consumption-by-type/; "Per Capita Meat Consumption in the United States in 2017 and 2018, by Type," Statista, released February 2019, https://www.statista.com /statistics/189222/average-meat-consumption-in-the-us-by-sort/.

36. "Broiler Meat Production Worldwide from 2012 to 2019," Statista, released April 2019, https://www.statista.com/statistics/237637/production-of-poultry-meat-worldwide-since-1990/; "Broiler Meat Production Worldwide in 2019, by Country," Statista, released April 2019, https://www.statista.com/statistics/237597 /leading-10-countries-worldwide-in-poultry-meat-production-in-2007/.

37. "Leading Egg Producing Countries Worldwide in 2018," Statista, released January 2020, https://www.statista.com/statistics/263971/top-10-countries-worldwide -in-egg-production/.

38. "Top U.S. States Based on Number of Laying Hens 2018," Statista, released January 2019, https://www.statista.com/statistics/196080/total-number-of-laying-hens-by -top-10-us-states/.

39. T. A. Ebeid, T. Suzuki, and T. Sugiyama, "High Ambient Temperature Influences Eggshell Quality and Calbindin-D28k Localization of Eggshell Gland and All Intestinal Segments of Laying Hens," abstract, *Poultry Science* 91, no. 9 (September 1, 2012):2282, https://doi.org/10.3382/ps.2011-01898.

40. St-Pierre, Cobanov, and Schnitkey, "Economic Losses," 52.

41. D. S. F. Lamarca et al., "Climate Change in Layer Poultry Farming: Impact of Heat Waves in Region of Bastos, Brazil," abstract, *Brazilian Journal of Poultry Science* 20, no. 4 (December 2018):657, https://doi.org/10.1590/1806-9061-2018-0750.

42. G. D. Vandana and V. Sejian, "Towards Identifying Climate Resilient Poultry Birds," abstract, *Journal of Dairy, Veterinary & Animal Research* 7 (May 16, 2018):84, https://doi.org/10.15406/jdvar.2018.07.00195.

43. Michael Graff, "Millions of Dead Chickens and Pigs Found in Hurricane Floods," *Guardian*, September 22, 2018, https://www.theguardian.com/environment/2018 /sep/21/hurricane-florence-flooding-north-carolina.

44. A. N. Hristov et al., "Climate Change Effects on Livestock in the Northeast US and Strategies for Adaptation," *Climatic Change* 146, no. 1 (January 1, 2018):39, https://doi.org/10.1007/s10584-017-2023-z.

45. A. R. Sharifi, P. Horst, and H. Simianer, "The Effect of Naked Neck Gene and Ambient Temperature and Their Interaction on Reproductive Traits of Heavy Broiler Dams," abstract, *Poultry Science* 89, no. 7 (July 2010):1360, https://doi.org/10.3382/ps.2009-00593.

46. Dubravka Skunca et al., "Life Cycle Assessment of the Chicken Meat Chain," abstract, *Journal of Cleaner Production* 184 (May 20, 2018):440, https://doi.org/10.1016/j.jclepro.2018.02.274.

47. Michael P. Richards and Erik Trinkaus, "Isotopic Evidence for the Diets of European Neanderthals and Early Modern Humans," abstract, *Proceedings of the National Academy of Sciences* 106, no. 38 (September 22, 2009):16034, https://doi.org/10.1073/pnas.0903821106.

48. David A. Kroodsma et al., "Tracking the Global Footprint of Fisheries," abstract, *Science* 359, no. 6378 (February 23, 2018):904, https://doi.org/10.1126/science.aao5646.

49. "Commercial Fishing Methods," Sustainable Fisheries, University of Washington, accessed August 28, 2019, https://sustainablefisheries-uw.org/seafood-101/commercial-fishing-methods/.

50. "Global Fish Production from 2002 to 2018," Statista, released January 2019, https://www.statista.com/statistics/264577/total-world-fish-production-since-2002/.

51. Yannick Rousseau et al., "Evolution of Global Marine Fishing Fleets and the Response of Fished Resources," abstract, *Proceedings of the National Academy of Sciences* 116, no. 25 (June 18, 2019):12238, https://doi.org/10.1073/pnas.1820344116.

52. *The State of World Fisheries and Aquaculture: Meeting the Sustainable Development Goals* (Rome: FAO, 2018), 72, http://www.fao.org/3/I9540EN/i9540en.pdf.

53. "Fish: Friend or Foe?," Nutrition Source, September 18, 2012, https://www.hsph.harvard.edu/nutritionsource/fish/.

54. Mathew Koll Roxy et al., "A Reduction in Marine Primary Productivity Driven by Rapid Warming over the Tropical Indian Ocean," *Geophysical Research Letters* 43, no. 2 (2016):826–33, https://doi.org/10.1002/2015GL066979.

55. Angus Atkinson et al., "Krill (*Euphausia superba*) Distribution Contracts Southward during Rapid Regional Warming," abstract, *Nature Climate Change* 9 (January 2019):142, https://www.nature.com/articles/s41558-018-0370-z.pdf.

56. Mary Papenfuss, "Mussels Cooked to Death in Their Shells in Unusual Heat on Northern California Shore," HuffPost, July 1, 2019, https://www.yahoo.com/huffpost/mussels-cooked-in-california-heat-bodega-climate-change-044813649.html.

57. Laurel Sheufelt, "The Population of Pollock under Climate Change as Determined by Age, Distribution, and Prey Energy Content," *ScienceBuzz* (blog), August 21, 2017, https://www.sciencebuzz.com/the-population-of-pollock-under-climate-change-as-determined-by-age-distribution-and-prey-energy-content/.

58. Denise Breitburg et al., "Declining Oxygen in the Global Ocean and Coastal Waters," *Science* 359, no. 6371 (January 5, 2018):eaam7240, https://doi.org/10.1126/science.aam7240.

59. C. D. Harvell et al., "Disease Epidemic and a Marine Heat Wave Are Associated with the Continental-Scale Collapse of a Pivotal Predator (*Pycnopodia helianthoides*)," *Science Advances* 5, no. 1 (January 1, 2019):1–8, https://doi.org/10.1126/sciadv.aau7042.

60. Craig Welch, "Climate Change May Spark Global 'Fish Wars,'" video, National Geographic News, June 14, 2018, https://news.nationalgeographic.com/2018/06/climate-change-drives-fish-wars-science-environment/.

61. Welch, "Climate Change."

62. Scott A. Doney et al., "Oceans and Marine Resources," in *Climate Change Impacts in the United States: The Third National Climate Assessment*, ed. J. M. Melillo, Terese (T.C.) Richmond, and G. W. Yohe (Washington, DC: US Global Change Research Program, 2014), 562, https://nca2014.globalchange.gov/report/regions/oceans.

63. Cosima S. Porteus et al., "Near-Future CO_2 Levels Impair the Olfactory System of a Marine Fish," abstract, *Nature Climate Change* 8, no. 8 (August 2018):737, https://doi.org/10.1038/s41558-018-0224-8.

64. David Bushek and Susan E. Ford, "Anthropogenic Impacts on an Oyster Metapopulation: Pathogen Introduction, Climate Change and Responses to Natural Selection," abstract, *Elementa: Science of the Anthropocene* 4 (August 18, 2016):1, https://doi.org/10.12952/journal.elementa.000119.

65. Reem Deeb et al., "Impact of Climate Change on *Vibrio vulnificus* Abundance and Exposure Risk," abstract, *Estuaries and Coasts: Journal of the Estuarine Research Federation* 41, no. 8 (December 2018):2289, https://doi.org/10.1007/s12237-018-0424-5.

66. "Swimming Upstream: Freshwater Fish in a Warming World," National Wildlife Federation, 2013, 16, https://www.nwf.org/~/media/PDFs/Global-Warming/Reports/NWF-Swimming%20Upstream-082813-B.ashx.

67. Timothy J. Bartley et al., "Food Web Rewiring in a Changing World," *Nature Ecology & Evolution* 3, no. 3 (March 2019):348, https://doi.org/10.1038/s41559-018-0772-3.

68. Tim Searchinger et al., *Creating a Sustainable Food Future: Synthesis Report* (World Resources Report, 2018), 40, https://wrr-food.wri.org/sites/default/files/2019-07/creating-sustainable-food-future_2_5.pdf.

69. Robert W. R. Parker et al., "Fuel Use and Greenhouse Gas Emissions of World Fisheries," *Nature Climate Change* 8, no. 4 (April 2018):335, https://doi.org/10.1038/s41558-018-0117-x.

70. Searchinger et al., *Creating a Sustainable Food Future*, 40–41.

71. Steven D. Gaines et al., "Improved Fisheries Management Could Offset Many Negative Effects of Climate Change," abstract, *Science Advances* 4, no. 8 (August 29, 2018):1, https://doi.org/10.1126/sciadv.aao1378.

72. *State of World Fisheries and Aquaculture*, FAO, 133.

73. Elizabeth Ashton, "The Impact of Shrimp Farming on Mangrove Ecosystems," *CAB Reviews: Perspectives in Agriculture, Veterinary Science, Nutrition and Natural Resources* 3 (2008):1–12, https://doi.org/10.1079/PAVSNNR20083003.

74. Emily Waltz, "First Genetically Engineered Salmon Sold in Canada," *Scientific American*, August 7, 2017, https://www.scientificamerican.com/article/first-genetically-engineered-salmon-sold-in-canada/.

75. Zoë A. Doubleday et al., "Global Proliferation of Cephalopods," *Current Biology* 26, no. 10 (May 2016):R406, https://doi.org/10.1016/j.cub.2016.04.002.

Grains, Starches, and Other Sides

1. Food and Agriculture Organization (FAO) of the United Nations, "Land under Cereal Production," World Bank, accessed March 6, 2020, https://data.worldbank.org/indicator/AG.LND.CREL.HA.

2. Hannah Ritchie and Max Roser, "Crop Yields," Our World in Data, October 17, 2013, https://ourworldindata.org/crop-yields.

3. Joseph M. Awika, "Major Cereal Grains Production and Use around the World," in *Advances in Cereal Science: Implications to Food Processing and Health Promotion*, ed. Joseph M. Awika, Vieno Piironen, and Scott Bean (Washington, DC: American Chemical Society, 2011), 2, https://doi.org/10.1021/bk-2011-1089.ch001.

4. Sumithra Muthayya et al., "An Overview of Global Rice Production, Supply, Trade, and Consumption," abstract, *Annals of the New York Academy of Sciences* 1324, no. 1 (September 2014):7, https://doi.org/10.1111/nyas.12540.

5. Ewen Callaway, "Domestication: The Birth of Rice," *Nature* 514, no. 7524 (October 2014):S58–59, https://doi.org/10.1038/514S58a.

6. Jae Young Choi et al., "The Complex Geography of Domestication of the African Rice *Oryza glaberrima*," abstract, *PLOS Genetics* 15, no. 3 (March 7, 2019):1, https://doi.org/10.1371/journal.pgen.1007414.

7. "Wild Rice," Wikipedia, last edited February 14, 2020, https://en.wikipedia.org/w/index.php?title=Wild_rice&oldid=930891440.

8. "Worldwide Production of Grain in 2018/19, by Type," Statista, released 2019, https://www.statista.com/statistics/263977/world-grain-production-by-type/.

9. "Rice Sector at a Glance," US Department of Agriculture Economic Research Service (USDA ERS), updated August 20, 2019, https://www.ers.usda.gov/topics/crops/rice/rice-sector-at-a-glance/#production; "Total U.S. Rice Production Value from 2000 to 2018," Statista, released April 2019, https://www.statista.com/statistics/190474/total-us-rice-production-value-from-2000/.

10. "Rice Sector," USDA ERS.

11. Suzanne K Redfern, Nadine Azzu, and Jesie S Binamira, "Rice in Southeast Asia: Facing Risks and Vulnerabilities to Respond to Climate Change," in *Building Resilience for Adaptation to Climate Change in the Agriculture Sector: Proceedings*

of a Joint FAO/OECD Workshop, 23–24 April, 2012, ed. Alexandre Meybeck et al. (Rome: FAO, 2012), 297, http://www.fao.org/3/i3084e/i3084e18.pdf.

12. "Salinity Intrusion in a Changing Climate Scenario Will Hit Coastal Bangladesh Hard," World Bank, February 17, 2015, https://www.worldbank.org/en/news/fea ture/2015/02/17/salinity-intrusion-in-changing-climate-scenario-will-hit-coastal-bang ladesh-hard.

13. Terrance Chea, "Drought Hurts California Rice Harvest," *Spokesman-Review*, October 30, 2014, http://www.spokesman.com/stories/2014/oct/30/california-drought -hurting-rice-harvest/.

14. Chuang Zhao et al., "Temperature Increase Reduces Global Yields of Major Crops in Four Independent Estimates," abstract, *Proceedings of the National Academy of Sciences* 114, no. 35 (August 29, 2017):9326, https://doi.org/10.1073 /pnas.1701762114.

15. Tetsuo Satake and Shouichi Yoshida, "High Temperature-Induced Sterility in Indica Rices at Flowering," *Japanese Journal of Crop Science* 47, no. 1 (1978):6–17, https://doi.org/10.1626/jcs.47.6; Shaobing Peng et al., "Rice Yields Decline with Higher Night Temperature from Global Warming," abstract, *Proceedings of the National Academy of Sciences* 101, no. 27 (July 6, 2004):9971, https://doi.org /10.1073/pnas.0403720101.

16. Chunwu Zhu et al., "Carbon Dioxide (CO_2) Levels This Century Will Alter the Protein, Micronutrients, and Vitamin Content of Rice Grains with Potential Health Consequences for the Poorest Rice-Dependent Countries," abstract, *Science Advances* 4, no. 5 (May 1, 2018):1, https://doi.org/10.1126/sciadv.aaq1012.

17. Marielle Saunois et al., "The Global Methane Budget 2000–2012," *Earth System Science Data* 8, no. 2 (December 12, 2016):711, https://doi.org/10.5194/essd -8-697-2016.

18. "IRRI at a Glance," International Rice Research Institute, accessed June 27, 2019, https://www.irri.org/irri-glance.

19. "About Us," Wheat and Rice Center for Heat Resilience, accessed February 22, 2019, http://wrchr.org/.

20. "About Us," California Cooperative Rice Research Foundation, Rice Experiment Station, accessed March 7, 2020, https://www.crrf.org/ccrrf_res_1-7-2020_006 .htm.

21. "Welcome to SRI-Rice Online!," System of Rice Intensification International Network and Resources Center, accessed May 7, 2019, http://sri.ciifad.cornell.edu /index.html.

22. Benjamin R. K. Runkle et al., "Methane Emission Reductions from the Alternate Wetting and Drying of Rice Fields Detected Using the Eddy Covariance Method," abstract, *Environmental Science & Technology* 53, no. 2 (January 15, 2019):671, https://doi.org/10.1021/acs.est.8b05535.

23. Yu Jiang et al., "Acclimation of Methane Emissions from Rice Paddy Fields to Straw Addition," abstract, *Science Advances* 5, no. 1 (January 1, 2019):1, https:// doi.org/10.1126/sciadv.aau9038.

24. Mary Alexander, "Sustainable Rice Farmers Sell First-Ever Carbon Credits to Microsoft," Microsoft Green Blog, May 10, 2018, https://blogs.microsoft.com/green/2018/05/10/sustainable-rice-farmers-sell-first-ever-carbon-credits-to-microsoft/.

25. M. L. Reba et al., "Aquifer Depletion in the Lower Mississippi River Basin: Challenges and Solutions," *Journal of Contemporary Water Research & Education* 162, no. 1 (2017):134, https://doi.org/10.1111/j.1936-704X.2017.03264.x.

26. Awika, "Major Cereal Grains," 5.

27. "Cereal Grains: The Structure & Uses of Wheat," ScienceAid, accessed April 8, 2019, https://scienceaid.net/Economic_Botany.

28. K. Kris Hirst, "Wheat Domestication: The History and Origins of Bread and Durum Wheat," ThoughtCo, updated July 9, 2019, https://www.thoughtco.com/wheat-domestication-the-history-170669.

29. S. Asseng et al., "Rising Temperatures Reduce Global Wheat Production," abstract, *Nature Climate Change* 5, no. 2 (February 2015):143, https://doi.org/10.1038/nclimate2470.

30. Miroslav Trnka et al., "Mitigation Efforts Will Not Fully Alleviate the Increase in Water Scarcity Occurrence Probability in Wheat-Producing Areas," abstract, *Science Advances* 5, no. 9 (September 1, 2019):1, https://doi.org/10.1126/sciadv.aau2406.

31. Andrew Crane-Droesch et al., "Climate Change and Agricultural Risk Management into the 21st Century," USDA ERS ERR-266, July 2019, https://www.ers.usda.gov/webdocs/publications/93547/err-266.pdf?v=9932.1.

32. Curtis A. Deutsch et al., "Increase in Crop Losses to Insect Pests in a Warming Climate," *Science* 361, no. 6405 (August 31, 2018):917, https://doi.org/10.1126/science.aat3466.

33. Peter Juroszek and Andreas von Tiedemann, "Climate Change and Potential Future Risks through Wheat Diseases: A Review," *European Journal of Plant Pathology* 136, no. 1 (May 1, 2013):25, https://doi.org/10.1007/s10658-012-0144-9.

34. Man Li, "Climate Change to Adversely Impact Grain Production in China by 2030," International Food Policy Research Institute, February 13, 2018, http://www.ifpri.org/blog/climate-change-adversely-impact-grain-production-china-2030.

35. James R. Hunt et al., "Early Sowing Systems Can Boost Australian Wheat Yields despite Recent Climate Change," abstract, *Nature Climate Change* 9, no. 3 (March 2019):244. https://doi.org/10.1038/s41558-019-0417-9.

36. Yasunari Ogihara, Shigeo Takumi, and Hirokazu Handa, eds., *Advances in Wheat Genetics: From Genome to Field: Proceedings of the 12th International Wheat Genetics Symposium* (Tokyo: SpringerOpen, 2015), https://link.springer.com/content/pdf/10.1007%2F978-4-431-55675-6.pdf.

37. Jessie Szalay, "Potatoes: Health Benefits, Risks & Nutrition Facts," Live Science, accessed March 7, 2020, https://www.livescience.com/45838-potato-nutrition.html.

38. "Potatoes: 2017 Summary," USDA National Agriculture Statistics Service, September 2018, https://www.nass.usda.gov/Publications/Todays_Reports/reports/pots0918.pdf.

39. "Potatoes," Agricultural Marketing Resource Center (AMRC), revised October 2018, https://www.agmrc.org/commodities-products/vegetables/potatoes.

40. "Potatoes," AMRC.

41. "Potatoes: 2017 Summary," USDA National Agriculture Statistics Service.

42. "Potatoes," AMRC.

43. Erik Stokstad, "The New Potato," abstract, *Science* 363, no. 6427 (February 8, 2019):574, https://doi.org/10.1126/science.363.6427.574.

44. "Potato Production Worldwide in 2018, by Leading Country," Statista, released February 2020, https://www.statista.com/statistics/382192/global-potato-production-by-country/.

45. "Diffusion," International Year of the Potato, FAO, 2008, http://www.fao.org/potato-2008/en/potato/diffusion.html.

46. "Potato Faces Up to Climate Change Challenges," *International Potato Center* (blog), April 2, 2013, https://cipotato.org/blog/potato-faces-up-to-climate-change-challenges/.

47. Tom Leavitt et al., eds., "Recipe for Disaster: How Climate Change Threatens British-Grown Fruit and Veg," Priestley International Centre for Climate, Leeds University, accessed March 7, 2020, 28, https://static1.squarespace.com/static/58b40fe1be65940cc4889d33/t/5c5bffc4652dea319f39bf6e/1549533170231/RecipeDisaster Report_Web_compressed.pdf.

48. Andre Daccache et al., "Climate Change and Land Suitability for Potato Production in England and Wales: Impacts and Adaptation," abstract, *Journal of Agricultural Science* 150, no. 2 (April 2012):161, https://doi.org/10.1017/S0021859611000839.

49. Keith Weatherhead and Nicholas Howden, "The Relationship between Land Use and Surface Water Resources in the UK," *Land Use Policy* 26 (December 1, 2009):S249, https://doi.org/10.1016/j.landusepol.2009.08.007.

50. Claudio O. Stöckle et al., "Assessment of Climate Change Impact on Eastern Washington Agriculture," *Climatic Change* 102, no. 1 (September 1, 2010):87, https://doi.org/10.1007/s10584-010-9851-4.

51. Rubí Raymundo et al., "Climate Change Impact on Global Potato Production," *European Journal of Agronomy* 100 (October 2018):96, https://doi.org/10.1016/j.eja.2017.11.008.

52. Naresh Kumar et al., "Assessment of Impact of Climate Change on Potato and Potential Adaptation Gains in the Indo-Gangetic Plains of India," abstract, *International Journal of Plant Production* 9 (2015):151, https://cgspace.cgiar.org/handle/10568/76568.

53. Rubí Raymundo et al., "Potato, Sweet Potato, and Yam Models for Climate Change: A Review," *Field Crops Research* 166 (September 2014):180, https://doi.org/10.1016/j.fcr.2014.06.017.

54. Stöckle et al., "Assessment of Climate Change Impact," 88.

55. Robert J. Hijmans, "The Effect of Climate Change on Global Potato Production," *American Journal of Potato Research* 80, no. 4 (July 2003):277, https://doi.org/10.1007/BF02855363.

56. "Potato Facts and Figures," *International Potato Center* (blog), accessed June 24, 2019, https://cipotato.org/crops/potato/potato-facts-and-figures/.

57. Stokstad, "New Potato," 577.

Dessert and Coffee

1. Adam Drewnowski et al., "Sweetness and Food Preference," abstract, *Journal of Nutrition* 142, no. 6 (June 2012):1142S, https://doi.org/10.3945/jn.111.149575.

2. Michael Krondl, *Sweet Invention: A History of Dessert* (Chicago: Chicago Review Press, 2011), 13.

3. Khalil Gibran Muhammad, "The Sugar That Saturates the American Diet Has a Barbaric History as the 'White Gold' That Fueled Slavery," *New York Times*, August 14, 2019, https://www.nytimes.com/interactive/2019/08/14/magazine/sugar-slave-trade-slavery.html.

4. "Sugar Cane Production Worldwide from 1965 to 2018," Statista, released January 2020, https://www.statista.com/statistics/249604/sugar-cane-production-worldwide/.

5. "Leading Sugar Cane Producers Worldwide in 2017, Based on Production Volume," Statista, released January 2019, https://www.statista.com/statistics/267865/principal-sugar-cane-producers-worldwide/.

6. Diane Huntrods and Vikram Koundinya, "Sugarcane Profile," Agricultural Marketing Resource Center, updated August 2018, https://www.agmrc.org/commodities-products/grains-oilseeds/sugarcane-profile.

7. "Leading Sugar Beet Producers Worldwide in 2017, Based on Production Volume," Statista, released January 2019, https://www.statista.com/statistics/264670/top-sugar-beet-producers-worldwide-by-volume/.

8. "Cyclone Larry Lashes Northeastern Queensland," US Department of Agriculture (USDA) Foreign Agricultural Service, March 23, 2006, https://ipad.fas.usda.gov/highlights/2006/03/australia_23mar2006/.

9. Duli Zhao and Yang-Rui Li, "Climate Change and Sugarcane Production: Potential Impact and Mitigation Strategies," *International Journal of Agronomy* (October 22, 2015):3, https://doi.org/10.1155/2015/547386.

10. P. D. Jones et al., "Future Climate Impact on the Productivity of Sugar Beet (*Beta vulgaris* L.) in Europe," abstract, *Climatic Change* 58, no. 1 (May 1, 2003):93, https://doi.org/10.1023/A:1023420102432.

11. "Retail Consumption of Chocolate Confectionery Worldwide from 2012/13 to 2018/19," Statista, released November 2015, https://www.statista.com/statistics/238849/global-chocolate-consumption/.

12. Gregory R. Ziegler, "Chocolate," in *The Encyclopedia of Food and Culture*, vol. 1, ed. Solomon H. Katz and William Woys Weaver (New York: Scribner's, 2003), 400, http://1.droppdf.com/files/q3gPk/encyclopedia-of-food-and-culture-volume-1.pdf.

13. Götz Schroth et al., "Vulnerability to Climate Change of Cocoa in West Africa: Patterns, Opportunities and Limits to Adaptation," *Science of the Total Environment* 556 (June 15, 2016):232, https://doi.org/10.1016/j.scitotenv.2016.03.024.

14. *True Price of Cocoa from the Ivory Coast*, joint report IDH and True Price, April 11, 2016, 6, https://issuu.com/idhsustainabletradeinitiative/docs/tp_cocoa_7.2__complete__web.

15. Johanna Jacobi et al., "Agroecosystem Resilience and Farmers' Perceptions of Climate Change Impacts on Cocoa Farms in Alto Beni, Bolivia," *Renewable Agriculture and Food Systems* 30, no. 2 (June 2015):172, 180, https://doi.org/10.1017/S174217051300029X.

16. "Size of the Chocolate Confectionery Market Worldwide from 2017 to 2026," Statista, released January 2018, https://www.statista.com/statistics/983554/global-chocolate-confectionery-market-size/.

17. "Cocoa Production Worldwide from 1980/81 to 2018/19," Statista, released May 2019, https://www.statista.com/statistics/262620/global-cocoa-production/.

18. "Per Capita Chocolate Consumption Worldwide in 2017, by Country," Statista, released March 2018, https://www.statista.com/statistics/819288/worldwide-chocolate-consumption-by-country/.

19. Michon Scott, "Climate & Chocolate," National Oceanic and Atmospheric Administration, February 10, 2016, https://www.climate.gov/news-features/climate-and/climate-chocolate.

20. Schroth et al., "Vulnerability to Climate Change," 231.

21. Lauranne Gateau-Rey et al., "Climate Change Could Threaten Cocoa Production: Effects of 2015–16 El Niño-Related Drought on Cocoa Agroforests in Bahia, Brazil," abstract, *PLOS ONE* 13, no. 7 (July 10, 2018):1, https://doi.org/10.1371/journal.pone.0200454.

22. Sania Ortega Andrade et al., "Climate Change and the Risk of Spread of the Fungus from the High Mortality of Theobroma Cocoa in Latin America," abstract, *Neotropical Biodiversity* 3, no. 1 (January 1, 2017):30, https://doi.org/10.1080/23766808.2016.1266072.

23. P. Läderach et al., "Predicting the Future Climatic Suitability for Cocoa Farming of the World's Leading Producer Countries, Ghana and Côte d'Ivoire," *Climatic Change* 119, no. 3 (August 1, 2013):852, https://doi.org/10.1007/s10584-013-0774-8.

24. J. Beer et al., "Shade Management in Coffee and Cacao Plantations," *Agroforestry Systems* 38, no. 1 (July 1, 1997):139–64, https://doi.org/10.1023/A:1005956528316.

25. Joanne Silberner, "How Chocolate Can Save the Planet," National Public Radio, November 19, 2007, https://www.npr.org/templates/story/story.php?storyId=16354380.

26. "About ICQC,R," International Cocoa Quarantine Center, accessed March 9, 2020, http://www.icgd.reading.ac.uk/icqc/about.php.

27. Andrew S. Fister et al., "Transient Expression of CRISPR/Cas9 Machinery Targeting TcNPR3 Enhances Defense Response in *Theobroma cacao*," abstract, *Frontiers in Plant Science* 9 (2018):1, https://doi.org/10.3389/fpls.2018.00268.

28. "International Cocoa Collection (IC3)," Tropical Agricultural Research and Higher Education Center (CATIE), accessed August 5, 2019, https://www.catie.ac.cr/en/products-and-services/collections-and-germplasm-banks/international-cocoa-collection.html.

29. "Vision & Mission," World Cocoa Foundation, accessed September 5, 2019, https://www.worldcocoafoundation.org/about-wcf/vision-mission/.

30. Tannis Thorlakson, "A Move beyond Sustainability Certification: The Evolution of the Chocolate Industry's Sustainable Sourcing Practices," abstract, *Business Strategy and the Environment* 27, no. 8 (2018):1653, https://doi.org/10.1002/bse.2230.

31. "Climate Action Position Statement," Mars, accessed September 6, 2019, https://www.mars.com/about/policies-and-practices/climate-action-position-statement.

32. Natasha Bach, "The World's Largest Chocolate Maker Is Committing $1 Billion to Fight Climate Change," *Fortune*, September 6, 2017, https://fortune.com/2017/09/06/mars-pledge-one-billion-fight-climate-change/.

33. "Ice Cream Sales & Trends," International Dairy Foods Association, accessed March 10, 2019, https://www.idfa.org/news-views/media-kits/ice-cream/ice-cream-sales-trends.

34. Khushboo Sheth, "Top Milk Producing Countries in the World," WorldAtlas, updated January 10, 2018, https://www.worldatlas.com/articles/top-cows-milk-producing-countries-in-the-world.html.

35. "Table 1. State Summary Highlights, 2017," in *Census of Agriculture: 2017*, USDA National Agricultural Statistics Service, last modified November 13, 2019, https://www.nass.usda.gov/Publications/AgCensus/2017/Full_Report/Volume_1,_Chapter_2_US_State_Level/st99_2_0001_0001.pdf.

36. James M. MacDonald et al., "Profits, Costs, and the Changing Structure of Dairy Farming," USDA Economic Research Service, September 1, 2007, 2, https://doi.org/10.2139/ssrn.1084458.

37. "Dairy Cattle and Milk Production," in *2017 Census of Agriculture Highlights*, USDA National Agricultural Statistics Service, September 4, 2019, https://www.nass.usda.gov/Publications/Highlights/2019/2017Census_DairyCattle_and_Milk_Production.pdf; "2018 US Dairy Statistics: 10-Year Changes by State," Progressive Dairy, accessed November 18, 2019, https://www.progressivepublish.com/downloads/2019/general/2018-pd-stats-highres.pdf, 2.

38. "Number of Milk Cows Worldwide by Country 2019," Statista, released December 2019, https://www.statista.com/statistics/869885/global-number-milk-cows-by-country/; Atli Arnarson, "Milk 101: Nutrition Facts and Health Effects," Healthline, March 25, 2019, https://www.healthline.com/nutrition/foods/milk.

39. Stefano Gerosa and Jakob Skoet, "Milk Availability: Trends in Production and Demand and Medium-Term Outlook" (Agricultural Development Economics Division [ESA] working paper no. 12–01, Food and Agriculture Organization of the United Nations [FAO], February 2012), 5, http://www.fao.org/3/an450e/an450e00.pdf.

40. Nikos Alexandratos and Jelle Bruinsma, "World Agriculture towards 2030/2050: The 2012 Revision" (ESA working paper no. 12–03, FAO, June 2012), 64, http://www.fao.org/3/a-ap106e.pdf; Phillip K. Thornton, "Livestock Production: Recent Trends, Future Prospects," *Philosophical Transactions: Biological Sciences* 365, no. 1554 (2010):2854, https://doi.org/10.1098/rstb.2010.0134.

41. Zhaohai Bai et al., "Global Environmental Costs of China's Thirst for Milk," abstract, *Global Change Biology* 24, no. 5 (2018):2198, https://doi.org/10.1111/gcb.14047.

42. Veronique Greenwood, "How Did Milk Become a Staple Food?," BBC Future, July 6, 2015, http://www.bbc.com/future/story/20150706-how-did-milk-become-a-staple-food.

43. Jeanine Bentley, "Trends in U.S. per Capita Consumption of Dairy Products, 1970–2012," USDA Economic Research Service, accessed March 16, 2020, https://www.ers.usda.gov/amber-waves/2014/june/trends-in-us-per-capita-consumption-of-dairy-products-1970-2012/.

44. Greg Thoma et al., "Greenhouse Gas Emissions from Milk Production and Consumption in the United States: A Cradle-to-Grave Life Cycle Assessment circa 2008," abstract, *International Dairy Journal* 31 (April 1, 2013):S3, https://doi.org/10.1016/j.idairyj.2012.08.013.

45. FAO and Global Dairy Platform, "Figure 10. Trends in Emission Intensity of Milk by Region (2005, 2010 and 2015)," *Climate Change and the Global Dairy Cattle Sector: The Role of the Dairy Sector in a Low-Carbon Future* (Rome: FAO and Global Dairy Platform, 2019), 25, http://www.fao.org/3/CA2929EN/ca2929en.pdf.

46. Kelsey Gee, "America's Dairy Farmers Dump 43 Million Gal. of Excess Milk," *Wall Street Journal*, updated October 12, 2016, https://www.wsj.com/articles/americas-dairy-farmers-dump-43-million-gallons-of-excess-milk-1476284353.

47. Rick Barrett, "Whipsawed by Low Milk Prices, More Dairy Farmers Call It Quits," *Milwaukee Journal Sentinel*, May 16, 2019, https://www.jsonline.com/story/news/special-reports/dairy-crisis/2019/05/16/whipsawed-low-milk-prices-more-wiconsin-dairy-farmers-call-quits/3304074002/.

48. P. Gowda et al. "Agriculture and Rural Communities," in *Impacts, Risks, and Adaptation in the United States: Fourth National Climate Assessment*, vol. 2, ed. D. R. Reidmiller et al. (Washington, DC: US Global Change Research Program, 2018), 398, https://doi.org/10.7930/NCA4.2018.CH10.

49. "Drinking Water for Dairy Cattle, Part 1," Dairy Herd Management, May 23, 2011, https://www.dairyherd.com/article/drinking-water-dairy-cattle-part-1; Larry E. Chase, "Climate Change Impacts on Dairy Cattle," University of Vermont, Burlington, https://www.uvm.edu/vtvegandberry/ClimateChangeAgricultureFactsheets.html.

50. Nigel Key and Stacy Sneeringer, "Potential Effects of Climate Change on the Productivity of U.S. Dairies," *American Journal of Agricultural Economics* 96, no. 4 (July 2014):1136–56, https://doi.org/10.1093/ajae/aau002; A. N. Hristov et al., "Climate Change Effects on Livestock in the Northeast US and Strategies for

Adaptation," *Climatic Change* 146, no. 1 (January 1, 2018):39, https://doi.org/10
.1007/s10584-017-2023-z.

51. A. Nardone et al., "Effects of Climate Changes on Animal Production and Sustainability of Livestock Systems," *Livestock Science* 130, no. 1–3 (May 2010):57–69, https://doi.org/10.1016/j.livsci.2010.02.011; Hristov et al., "Climate Change Effects," 34.

52. Key and Sneeringer, "Potential Effects," 1154.

53. Peter Wright, Department of Animal Science, College of Agriculture and Life Sciences, Cornell University, personal communication, March 29, 2019.

54. X. A. Ortiz et al., "Evaluation of Conductive Cooling of Lactating Dairy Cows under Controlled Environmental Conditions," abstract, *Journal of Dairy Science* 98, no. 3 (March 1, 2015):1759, https://doi.org/10.3168/jds.2014-8583; "Dual Chamber Cow Waterbeds," Advanced Comfort Technology, http://www.dccwaterbeds.com/; "Cooling Systems for Georgia Dairy Cattle," University of Georgia Extension, accessed April 10, 2019, http://extension.uga.edu/publications/detail.html?number=B1172&title=Cooling%20Systems%20for%20Georgia%20Dairy%20Cattle.

55. Chase, "Climate Change Impacts," 21.

56. S. R. Davis, R. J. Spelman, and M. D. Littlejohn, "Breeding Heat Tolerant Dairy Cattle: The Case for Introgression of the 'Slick' Prolactin Receptor Variant into Dairy Breeds," abstract, *Journal of Animal Science* 95, no. 4 (2017):1788, https://doi.org/10.2527/jas2016.0956.

57. Heather Huson et al., "Genome-Wide Association Study and Ancestral Origins of the Slick-Hair Coat in Tropically Adapted Cattle," abstract, *Frontiers in Genetics* 5 (April 29, 2014):1, https://doi.org/10.3389/fgene.2014.00101.

58. S. Dikmen et al, "The SLICK Hair Locus Derived from Senepol Cattle Confers Thermotolerance to Intensively Managed Lactating Holstein Cows," *Journal of Dairy Science* 97, no. 9 (September 2014):5508–20, https://doi.org/10.3168/jds.2014-8087.

59. R. John Wallace et al., "A Heritable Subset of the Core Rumen Microbiome Dictates Dairy Cow Productivity and Emissions," *Science Advances* 5, no. 7 (July 1, 2019):1–12, https://doi.org/10.1126/sciadv.aav8391.

60. A. N. Hristov et al., "Mitigation of Methane and Nitrous Oxide Emissions from Animal Operations: I. A Review of Enteric Methane Mitigation Options," *Journal of Animal Science* 91 (2013):5045–69, https://academic.oup.com/jas/article/91/11/5045/4731308.

61. "3-NOP Set to Block Methane in Rumens," Rural News Group, July 11, 2018, https://www.ruralnewsgroup.co.nz/dairy-news/dairy-farm-health/3-nop-set-to-block-methane-in-rumens; D. Van Wesemael et al., "Reducing Enteric Methane Emissions from Dairy Cattle: Two Ways to Supplement 3-Nitrooxypropanol," *Journal of Dairy Science* 102, no. 2 (February 1, 2019):1780–87, https://doi.org/10.3168/jds.2018-14534.

62. Breanna M. Roque et al., "Inclusion of *Asparagopsis armata* in Lactating Dairy Cows' Diet Reduces Enteric Methane Emission by over 50 Percent," abstract, *Journal of Cleaner Production* 234 (October 10, 2019):132, https://doi.org/10.1016/j.jclepro.2019.06.193.

63. Rebecca Rupp, "How Garlic May Save the World," *National Geographic*, April 24, 2014, https://www.nationalgeographic.com/culture/food/the-plate/2014/04/24/how-garlic-may-save-the-world/.

64. Nathan Hurst, "Why Anaerobic Digestion Is Becoming the Next Big Renewable Energy Source," *Smithsonian*, November 3, 2016, https://www.smithsonianmag.com/innovation/why-anaerobic-digestion-becoming-next-big-renewable-energy-source-180960992/.

65. Elizabeth Newbold, "Anaerobic Digesters," Cornell Small Farms Program, June 11, 2013, https://smallfarms.cornell.edu/2013/06/11/anaerobic-digesters/.

66. "Methane Gas Recovery," Wisconsin Public Service, accessed March 12, 2019, https://accel.wisconsinpublicservice.com/business/methane_gas.aspx.

67. "Anaerobic Digester Facts and Trends," Environmental Protection Agency (EPA) AgSTAR, 2019, https://www.epa.gov/agstar/agstar-data-and-trends#adfacts.

68. Albert Morales, "Why Don't We Hear Much about Anaerobic Digestion in the U.S.?," Renewable Energy World, September 18, 2007, https://www.renewableenergyworld.com/articles/2007/09/why-dont-we-hear-much-about-anaerobic-digestion-in-the-u-s-49970.html; "Funding On-Farm Anaerobic Digestion," EPA AgSTAR, https://www.epa.gov/sites/production/files/2014-12/documents/funding_digestion.pdf.

69. Tim McAllister et al., "Greenhouse Gases in Animal Agriculture: Finding a Balance between Food Production and Emissions," *Animal Feed Science and Technology* (June 23, 2011):166–67; Peter Wright, Department of Animal Science, Cornell University, personal communication, March 29, 2019.

70. "AgSTAR Data and Trends," EPA AgSTAR, October 2019, https://www.epa.gov/agstar/agstar-data-and-trends.

71. Dairy Cares, "Cow Pies in the Sky: The Reality on Cows and Climate Goals," California Dairy Research Foundation, March 8, 2019, http://cdrf.org/2019/03/08/cow-pies-in-the-sky-the-reality-on-cows-and-climate-goals/.

72. Hurst, "Why Anaerobic Digestion."

73. C. Opio et al., *Greenhouse Gas Emissions from Ruminant Supply Chains: A Global Life Cycle Assessment* (Rome: FAO, 2013), 45, http://www.fao.org/3/i3461e/i3461e.pdf.

74. M. Vermorel, "Yearly Methane Emissions of Digestive Origin by Sheep, Goats and Equines in France: Variations with Physiological Stage and Production Type," *Productions Animales* 10 (2) (1995):153–61; G. Zervas and E. Tsiplakou, "An Assessment of GHG Emissions from Small Ruminants in Comparison with GHG Emissions from Large Ruminants and Monogastric Livestock," *Atmospheric Environment* 49 (March 1, 2012):13–23, https://doi.org/10.1016/j.atmosenv.2011.11.039.

75. Nissim Silanikove and Nazan Koluman (Darcan), "Impact of Climate Change on the Dairy Industry in Temperate Zones: Predications on the Overall Negative Impact and on the Positive Role of Dairy Goats in Adaptation to Earth Warming," *Small Ruminant Research* 123, no. 1 (January 1, 2015):27–34, https://doi.org/10.1016/j.smallrumres.2014.11.005.

76. Stefano Gerosa and Jakob Skoet, "Milk Availability: Trends in Production and Demand and Medium-Term Outlook" (ESA working paper 289000, FAO, 2012), 21, https://ideas.repec.org/p/ags/faoaes/289000.html.

77. G. F. W. Haenlein, "Past, Present, and Future Perspectives of Small Ruminant Dairy Research," *Journal of Dairy Science* 84, no. 9 (September 2001):2099, https://doi.org/10.3168/jds.S0022-0302(01)74655-3.

78. Silanikove and Koluman, "Impact of Climate Change."

79. Haenlein, "Past, Present, and Future Perspectives," 2097.

80. "Dairy Goats," Agricultural Marketing Resource Center, revised December 2018, https://www.agmrc.org/commodities-products/livestock/goats/dairy-goats.

81. "U.S. Dairy Goat Operations," Veterinary Services, Centers for Epidemiology and Animal Health, USDA Animal and Plant Health Information Service, March 2012, https://www.aphis.usda.gov/animal_health/nahms/goats/downloads/goat09/Goat09_is_DairyGoatOps.pdf.

82. Mathew Attokaran, "Vanilla," in *Natural Food Flavors and Colorants* (Oxford: Wiley-Blackwell, 2011), 403, https://doi.org/10.1002/9780470959152.ch99.

83. Donovan S. Correll, "Vanilla: Its Botany, History, Cultivation and Economic Import," *Economic Botany* 7, no. 4 (1953):300, https://link.springer.com/article/10.1007/BF02930810.

84. "Vanilla," Observatory of Economic Complexity, accessed August 15, 2019, https://oec.world/en/profile/hs92/0905/.

85. "Why There Is a Worldwide Shortage of Vanilla," *Economist*, March 28, 2018, https://www.economist.com/the-economist-explains/2018/03/28/why-there-is-a-worldwide-shortage-of-vanilla.

86. Nethaji J. Gallage and Birger Lindberg Møller, "Vanilla: The Most Popular Flavour," in *Biotechnology of Natural Products*, ed. Wilfried Schwab, Bernd Markus Lange, and Matthias Wüst (Cham, Switzerland: Springer, 2018), 3, https://doi.org/10.1007/978-3-319-67903-7_1.

87. Nethaji J. Gallage and Birger Lindberg Møller, "Vanillin—Bioconversion and Bioengineering of the Most Popular Plant Flavor and Its De Novo Biosynthesis in the Vanilla Orchid," *Molecular Plant* 8, no. 1 (January 5, 2015):41, https://doi.org/10.1016/j.molp.2014.11.008; Jeff Gelski, "Vanilla Prices Slowly Drop as Crop Quality Improves," *Food Business News*, March 22, 2019, https://www.foodbusinessnews.net/articles/13570-vanilla-prices-slowly-drop-as-crop-quality-improves?v=preview.

88. Celia A. Harvey et al., "Extreme Vulnerability of Smallholder Farmers to Agricultural Risks and Climate Change in Madagascar," abstract, *Philosophical Transactions of the Royal Society B: Biological Sciences* 369, no. 1639 (April 5, 2014):1, https://doi.org/10.1098/rstb.2013.0089.

89. Chase Purdy, "A Cyclone in Madagascar Has Made Vanilla Four Times More Expensive," *Quartz*, August 22, 2019, https://qz.com/1059470/why-is-vanilla-so -expensive-look-to-weather-in-madagascar/.

90. Bernard Giraud, "The Livelihoods—Vanilla Project in Madagascar: Towards a New Relation between Smallholder Farmers & Companies?," *Livelihoods Funds*, June 13, 2018, http://www.livelihoods.eu/transforming-the-relationship -between-smallholder-famers-companies-an-example-with-the-livelihoods -vanilla-project/.

91. Danone, "Transforming the Vanilla Supply Chain in Madagascar," Medium, September 7, 2017, https://medium.com/@Danone/transforming-the-vanilla-su pply-chain-in-madagascar-99e98dcabe5f.

92. Nethaji and Møller, "Vanilla."

93. "Animal Food Labeling," *Code of Federal Regulations*, title 21, chapter I, sub-chapter E, part 501, accessed August 21, 2019, https://www.accessdata.fda.gov /scripts/cdrh/cfdocs/cfcfr/cfrsearch.cfm?fr=501.22.

94. Juliana dePaula and Adriana Farah, "Caffeine Consumption through Coffee: Content in the Beverage, Metabolism, Health Benefits and Risks," *Beverages* 5, no. 2 (June 2019):37, https://doi.org/10.3390/beverages5020037.

95. "Coffee Production Worldwide from 2003/04 to 2018/19," Statista, released January 2020, https://www.statista.com/statistics/263311/worldwide-production -of-coffee/; "World's Largest Coffee Producing Countries in 2018," Statista, released January 2019, https://www.statista.com/statistics/277137/world-coffee-pro duction-by-leading-countries/.

96. "What Is Coffee?," National Coffee Association USA, accessed August 8, 2019, http://www.ncausa.org/About-Coffee/What-is-Coffee.

97. Marcelo Teixeira, Phuong Nguyen, and Julia Cobb, "How Brazil and Vietnam Are Tightening Their Grip on the World's Coffee," *Reuters*, August 22, 2019, https://www.reuters.com/article/us-coffee-duopoly-insight-idUSKCN1VC079.

98. Corey Watts, "A Brewing Storm: The Climate Change Risks to Coffee," Climate Institute, August 29, 2016, 1, http://www.climateinstitute.org.au/coffee .html.

99. "NCA National Coffee Data Trends 2019," National Coffee Association USA, March 9, 2019, https://nationalcoffee.blog/2019/03/09/national-coffee-drinking -trends-2019/; "The Economic Impact of the Coffee Industry," National Coffee Association USA, accessed August 9, 2019, http://www.ncausa.org/Industry-Resour ces/Economic-Impact.

100. "Global Leading Coffee Consuming Countries in 2015," Statista, released September 2016, https://www.statista.com/statistics/277135/leading-countries-by -coffee-consumption/.

101. "Number of Starbucks Stores Worldwide from 2003 to 2018," Statista, released November 2019, https://www.statista.com/statistics/266465/number-of-starbucks -stores-worldwide/.

102. Watts, "Brewing Storm," 4.

103. A. C. W. Craparo et al., "*Coffea arabica* Yields Decline in Tanzania Due to Climate Change: Global Implications," abstract, *Agricultural and Forest Meteorology* 207 (July 15, 2015):1, https://doi.org/10.1016/j.agrformet.2015.03.005.

104. C. Gay et al., "Potential Impacts of Climate Change on Agriculture: A Case Study of Coffee Production in Veracruz, Mexico," abstract, *Climatic Change* 79, no. 3 (December 1, 2006):259, https://doi.org/10.1007/s10584-006-9066-x.

105. Christian Bunn et al., "A Bitter Cup: Climate Change Profile of Global Production of Arabica and Robusta Coffee," abstract, *Climatic Change* 129, no. 1 (March 1, 2015):89, https://doi.org/10.1007/s10584-014-1306-x.

106. Aaron P. Davis et al., "High Extinction Risk for Wild Coffee Species and Implications for Coffee Sector Sustainability," abstract, *Science Advances* 5, no. 1 (January 1, 2019):1, https://doi.org/10.1126/sciadv.aav3473.

107. Fábio M. DaMatta et al., "Why Could the Coffee Crop Endure Climate Change and Global Warming to a Greater Extent Than Previously Estimated?," *Climatic Change* 152, no. 1 (January 2019):167–78, https://doi.org/10.1007/s105 84-018-2346-4.

108. Jacques Avelino et al., "The Coffee Rust Crises in Colombia and Central America (2008–2013): Impacts, Plausible Causes and Proposed Solutions," abstract, *Food Security* 7, no. 2 (April 1, 2015):303, https://doi.org/10.1007/s125 71-015-0446-9.

109. Watts, "Brewing Storm," 5.

110. Justin Worland, "Your Morning Cup of Coffee Is in Danger: Can the Industry Adapt in Time?," *Time*, June 21, 2018, https://time.com/5318245/coffee -industry-climate-change/.

111. "Shade-Grown Coffee," Wikipedia, edited December 10, 2019, https://en.wiki pedia.org/w/index.php?title=Shade-grown_coffee&oldid=930115080.

112. Justin Moat et al., "Resilience Potential of the Ethiopian Coffee Sector under Climate Change," abstract, *Nature Plants* 3, no. 7 (June 2017):1, https://doi.org /10.1038/nplants.2017.81.

113. Luc Cohen and Ivan Castro, "As Climate Change Threatens CentAm Coffee, a Cocoa Boom Is Born," *Reuters*, January 18, 2016, https://www.reuters.com/ar ticle/us-climatechange-cocoa-coffee-idUSKCN0UW1AV.

114. Jefferson Shriver, "Responding to the Climate Crisis through Crop Diversification," Daily Coffee News, October 26, 2015, https://dailycoffeenews.com/2015 /10/26/responding-to-the-climate-crisis-through-crop-diversification/.

115. Ally Coffee, "Coffee, Carbon, and Climate Change," Medium, May 28, 2019, https://medium.com/ally-coffee/coffee-carbon-and-climate-change-3bb7276 f8678.

116. "Tackling Climate Change," Starbucks Coffee Company, accessed February 17, 2020, https://www.starbucks.com/responsibility/environment/climate-change.

Farmers, Businesses, and Scientists

1. Zoe Willingham and Andy Green, "A Fair Deal for Farmers," Center for American Progress, May 7, 2019, https://www.americanprogress.org/issues/economy/reports/2019/05/07/469385/fair-deal-farmers/.

2. "Ag and Food Sectors and the Economy," US Department of Agriculture (USDA) Economic Research Service (ERS), updated September 20, 2019, https://www.ers.usda.gov/data-products/ag-and-food-statistics-charting-the-essentials/ag-and-food-sectors-and-the-economy.aspx.

3. "Most Farms Are Small, but Most Production Is on Large Farms," USDA ERS, updated November 27, 2019, http://www.ers.usda.gov/data-products/chart-gallery/gallery/chart-detail/?chartId=58288.

4. "Farm Household Income Forecast for 2019 and 2020," USDA ERS, updated February 5, 2020, https://www.ers.usda.gov/topics/farm-economy/farm-household-well-being/farm-household-income-forecast.

5. "Agricultural Safety," Center for Disease Control and Prevention (CDC), National Institute for Occupational Safety and Health, reviewed October 9, 2019, https://www.cdc.gov/niosh/topics/aginjury/default.html; Debbie Weingarten, "Why Are America's Farmers Killing Themselves?," Guardian, December 11, 2018, https://www.theguardian.com/us-news/2017/dec/06/why-are-americas-farmers-killing-themselves-in-record-numbers.

6. "Climate Smart Farming Story: Oechsner Farm," video produced by the Cornell Institute for Climate Change and Agriculture, November 24, 2015, http://climatesmartfarming.org/videos/oechsner-farm/.

7. "About," Food and Agriculture Organization of the United Nations (FAO) Global Alliance for Climate-Smart Agriculture, accessed July 11, 2019, http://www.fao.org/gacsa/about/en/; "Home," North American Climate Smart Agriculture Alliance, accessed July 11, 2019, https://www.nacsaa.net.

8. "Climate-Smart Agriculture," FAO, accessed May 15, 2019, http://www.fao.org/climate-smart-agriculture/en/.

9. "Climate Smart Farming Tools," Cornell University, Climate Smart Farming Program, accessed May 31, 2019, http://climatesmartfarming.org/tools/.

10. US Global Change Research Program, "Tools," U.S. Climate Resilience Toolkit, accessed June 6, 2019, https://toolkit.climate.gov/tools?f%5B0%5D=field_parent_topic%3A115.

11. P. Gowda et al., "Agriculture and Rural Communities," in Impacts, Risks, and Adaptation in the United States: Fourth National Climate Assessment, vol. 2, ed. D. R. Reidmiller et al. (Washington, DC: US Global Change Research Program, 2018), 401, https://doi.org/10.7930/NCA4.2018.CH10.

12. "Digital Farming Decisions and Insights to Maximize Every Acre," Climate Corporation, accessed July 8, 2019, https://climate.com/?gclid=CjwKCAjwo4vpBRB3EiwAoIieanreQ3lr16fgDQSPKjT-4ayhoDgeysRJ9ZcA4PoWr5WYGv2PPd2pjBoCDJcQAvD_BwE.

13. Alternative Farming Systems Information Center, "Organic Production/Organic Food: Information Access Tools," USDA, National Agricultural Library, accessed July 5, 2019, https://www.nal.usda.gov/afsic/organic-productionorganic-food-in formation-access-tools#define.

14. "Organic Agriculture," USDA ERS, updated October 9, 2019, https://www.ers.usda. gov/topics/natural-resources-environment/organic-agriculture/; Kristen Bialik and Kristi Walker, "Organic Farming Is on the Rise in the U.S.," Fact Tank, Pew Research Center, January 10, 2019, https://www.pewresearch.org/fact-tank/2019/01/10/orga nic-farming-is-on-the-rise-in-the-u-s/.

15. Hannah Ritchie, "Yields vs. Land Use: How the Green Revolution Enabled Us to Feed a Growing Population," Our World in Data, accessed November 10, 2019, https://ourworldindata.org/yields-vs-land-use-how-has-the-world-produced-eno ugh-food-for-a-growing-population; Alon Tal, "Making Conventional Agriculture Environmentally Friendly: Moving beyond the Glorification of Organic Agriculture and the Demonization of Conventional Agriculture," *Sustainability* 10, no. 4 (April 2018):1–17, https://doi.org/10.3390/su10041078.

16. John Dobberstein, "Census of Ag: Cover Crop Acres in U.S. Growing 8% Per Year," Cover Crop Strategies, April 16, 2019, https://www.covercropstrategies.com /articles/178-census-of-ag-cover-crop-acres-in-us-growing-8-per-year.

17. Roger Claassen et al., "Tillage Intensity and Conservation Cropping in the United States," USDA ERS, September 2018, https://www.ers.usda.gov/webdocs/publica tions/90201/eib-197.pdf?v=1783.8.

18. "Perennial Grain Crop Development," Land Institute, accessed July 8, 2019, https:// landinstitute.org/our-work/perennial-crops/.

19. Valérie Masson-Delmotte et al., eds., "Summary for Policymakers," in *Climate Change and Land: An IPCC Special Report on Climate Change, Desertification, Land Degradation, Sustainable Land Management, Food Security, and Greenhouse Gas Fluxes in Terrestrial Ecosystems*, ed. P. R. Shukla et al. (Geneva: Intergovernmental Panel on Climate Change, 2019), 10, https://www.ipcc.ch/site/assets/uploads /sites/4/2020/02/SPM_Updated-Jan20.pdf.

20. R. Lal et al., "Soil Carbon Sequestration to Mitigate Climate Change and Advance Food Security," abstract, *Soil Science* 172, no. 12 (December 2007):943, https://doi .org/10.1097/ss.0b013e31815cc498.

21. R. Lal, "Soil Carbon Sequestration Impacts on Global Climate Change and Food Security," abstract, *Science* 304, no. 5677 (June 11, 2004):1623, https://doi.org/10 .1126/science.1097396.

22. Sonja Vermeulen et al., "A Global Agenda for Collective Action on Soil Carbon," *Nature Sustainability* 2, no. 1 (January 2019):2–4, https://doi.org/10.1038/s41893 -018-0212-z.

23. "Danone Ecosystem Fund," Danone Ecosystem, accessed March 12, 2020, http:// ecosysteme.danone.com/.

24. "Project Gigaton," Walmart, accessed July 18, 2019, https://www.walmartsustain abilityhub.com/project-gigaton.

25. "About," Agricultural and Forestry Biodiversity Information Commons, accessed March 9, 2020, http://www.agrobiodiversity.org/about.

26. Katherine C. R. Baldock et al., "A Systems Approach Reveals Urban Pollinator Hotspots and Conservation Opportunities," abstract, *Nature Ecology & Evolution* 3, no. 3 (March 2019):363, https://doi.org/10.1038/s41559-018-0769-y.

27. Svetlana A. Chechetka et al., "Materially Engineered Artificial Pollinators," *Chem* 2, no. 2 (February 9, 2017):224–39, https://doi.org/10.1016/j.chempr.2017.01.008; Simon G. Potts et al., "Robotic Bees for Crop Pollination: Why Drones Cannot Replace Biodiversity," *Science of The Total Environment* 642 (November 2018): 665–67, https://doi.org/10.1016/j.scitotenv.2018.06.114.

28. *United States Summary and State Data, Volume 1: Geographic Area Series, Part 51, 2017 Census of Agriculture*, USDA National Agricultural Statistics Service, accessed March 12, 2020, 60, https://www.nass.usda.gov/Publications/AgCensus/2017/Full_Report/Volume_1,_Chapter_1_US/usv1.pdf.

29. Shen Ming Lee, *Hungry for Disruption: How Tech Innovations Will Nourish 10 Billion by 2050* (Potomac, MD: New Degree, 2019).

30. C. F. Nicholson et al., "An Economic and Environmental Comparison of Conventional and Controlled Environment Agriculture (CEA) Supply Chains for Leaf Lettuce to US Cities," in *Food Supply Chains in Cities: Modern Tools for Circularity and Sustainability*, ed. E. Aktas and Michael Bourlakis (London: Palgrave, 2020).

31. P. Gowda et al., "Agriculture and Rural Communities," in *Impacts, Risks, and Adaptation in the United States: Fourth National Climate Assessment*, vol. 2, ed. D. R. Reidmiller et al. (Washington, DC: US Global Change Research Program, 2018), 410–11, https://doi.org/10.7930/NCA4.2018.CH10.

32. Paul Heisey and Keith Fuglie, "Agricultural Research Investment and Policy Reform in High-Income Countries," USDA ERS, accessed March 12, 2020, 11, http://www.ers.usda.gov/publications/pub-details/?pubid=89113; "USDA ERS—Agricultural Research Funding in the Public and Private Sectors," updated September 2019, https://www.ers.usda.gov/data-products/agricultural-research-funding-in-the-public-and-private-sectors/.

33. Felix Richter, "Americans Spent $43 Billion on Video Games in 2018," Statista, June 11, 2019, https://www.statista.com/chart/9838/consumer-spending-on-video-games/.

34. Paul W. Fuglie and Keith O. Heisey, "Economic Returns to Public Agricultural Research," September 2007, USDA ERS, https://www.ers.usda.gov/webdocs/publications/42826/11496_eb10_1_.pdf?v=0.

35. "Strategic Vision," USDA Agriculture Research Service National Soil and Air Program, accessed July 15, 2019, https://www.ars.usda.gov/natural-resources-and-sustainable-agricultural-systems/soil-and-air/; "Climate Change," USDA National Resources Conservation Service, accessed July 15, 2019, https://www.nrcs.usda.gov/wps/portal/nrcs/main/national/climatechange/.

36. "Global Change and Climate Programs," USDA National Institute of Food and Agriculture, accessed July 9, 2019, https://nifa.usda.gov/program/global-change-and-climate-programs.

37. "About Us," USDA Climate Hubs, accessed July 9, 2019, https://www.climatehubs.usda.gov/about-us.

38. "Regional Climate Centers," National Oceanic and Atmospheric Administration, National Centers for Environmental Information, accessed July 9, 2019, https://www.ncdc.noaa.gov/customer-support/partnerships/regional-climate-centers.

39. "About USGCRP," US Global Change Research Program, accessed July 9, 2019, https://www.globalchange.gov/about.

40. Daniel Tobin et al., "Addressing Climate Change Impacts on Agriculture and Natural Resources: Barriers and Priorities for Land-Grant Universities in the Northeastern United States," abstract, *Weather, Climate, and Society* 9, no. 3 (May 24, 2017):591, https://doi.org/10.1175/WCAS-D-16-0106.1.

41. American Association for the Advancement of Science, "And Science's 2015 Breakthrough of the Year Is . . . ," *Science Magazine*, accessed November 13, 2019, https://www.sciencemag.org/news/2015/12/and-science-s-2015-breakthrough-year.

42. "A CRISPR Definition of Genetic Modification," *Nature Plants* 4, no. 5 (May 2018):233, https://doi.org/10.1038/s41477-018-0158-1.

43. Kunling Chen et al., "CRISPR/Cas Genome Editing and Precision Plant Breeding in Agriculture," *Annual Review of Plant Biology* 70, no. 1 (2019):667–97, https://doi.org/10.1146/annurev-arplant-050718-100049; Tian Wang, Hongyan Zhang, and Hongliang Zhu, "CRISPR Technology Is Revolutionizing the Improvement of Tomato and Other Fruit Crops," abstract, *Horticulture Research* 6, no. 1 (June 15, 2019):1, https://doi.org/10.1038/s41438-019-0159-x.

44. "Recent Trends in GE Adoption," USDA ERS, updated September 18, 2019, https://www.ers.usda.gov/data-products/adoption-of-genetically-engineered-crops-in-the-us/recent-trends-in-ge-adoption.aspx.

45. "Recent Trends in GE Adoption," USDA ERS.

46. Paul F. South et al., "Synthetic Glycolate Metabolism Pathways Stimulate Crop Growth and Productivity in the Field," abstract, *Science* 363, no. 6422 (January 4, 2019):1, https://doi.org/10.1126/science.aat9077.

47. Lewis H. Ziska et al., "Food Security and Climate Change: On the Potential to Adapt Global Crop Production by Active Selection to Rising Atmospheric Carbon Dioxide," *Proceedings of the Royal Society B: Biological Sciences* 279, no. 1745 (October 22, 2012):4097–105, https://doi.org/10.1098/rspb.2012.1005.

48. Alisdair R. Fernie and Jianbing Yan, "De Novo Domestication: An Alternative Route toward New Crops for the Future," abstract, *Molecular Plant* 12, no. 5 (May 6, 2019):615, https://doi.org/10.1016/j.molp.2019.03.016.

49. National Academies of Sciences, Engineering, and Medicine, *Genetically Engineered Crops: Experiences and Prospects* (Washington, DC: National Academies, 2016), https://doi.org/10.17226/23395.

50. Rhodora R. Aldemita and Randy A. Hautea, "Biotech Crop Planting Resumes High Adoption in 2016," abstract, *GM Crops & Food* 9, no. 1 (January 2, 2018):1, https://doi.org/10.1080/21645698.2018.1428166.

51. National Academies, *Genetically Engineered Crops*, 3.

52. I. Heap, "Weeds Resistant to EPSP Synthase Inhibitors (G/9)," International Survey of Herbicide Resistant Weeds, accessed May 15, 2019, http://www.weedscience .com/Summary/MOA.aspx?MOAID=12; Bruce E. Tabashnik, Thierry Brévault, and Yves Carrière, "Insect Resistance to *Bt* Crops: Lessons from the First Billion Acres," abstract, *Nature Biotechnology* 31, no. 6 (June 2013):510, https://doi.org /10.1038/nbt.2597.

53. "Stacked Traits in Biotech Crops," International Service for the Acquisition of Agri-biotech, updated March 2020, https://www.isaaa.org/resources/publications /pocketk/42/default.asp.

54. National Academies, "Genetically Engineered Crops," 91.

55. "National Bioengineered Food Disclosure Standard," *Federal Register*, December 21, 2018, https://www.federalregister.gov/documents/2018/12/21/2018-27283 /national-bioengineered-food-disclosure-standard.

56. "Genetically Modified Animal," Wikipedia, May 8, 2019, https://en.wikipedia .org/w/index.php?title=Genetically_modified_animal&oldid=896039418.

57. Gary Walsh, "Therapeutic Insulins and Their Large-Scale Manufacture," *Applied Microbiology and Biotechnology* 67, no. 2 (April 1, 2005):151–59, https://doi.org /10.1007/s00253-004-1809-x.

What We Can Do

1. *Climate Literacy: The Essential Principles of Climate Literacy*, second version (Washington, DC: US Global Change Research Program, March 2009), https:// downloads.globalchange.gov/Literacy/climate_literacy_highres_english.pdf.

2. Katharine Hayhoe, "The Most Important Thing You Can Do to Fight Climate Change: Talk about It," TED Women 2018, transcript accessed July 11, 2019, https://www.ted.com/talks/katharine_hayhoe_the_most_important_thing_you _can_do_to_fight_climate_change_talk_about_it/transcript.

3. Jennifer Marlon et al., "Yale Climate Opinion Maps 2019," Yale Program on Climate Change Communication, September 17, 2019, https://climatecommunication .yale.edu/visualizations-data/ycom-us/.

4. Matthew H. Goldberg et al., "Discussing Global Warming Leads to Greater Acceptance of Climate Science," abstract, *Proceedings of the National Academy of Sciences*, 116, no. 30 (July 3, 2019):14804, https://doi.org/10.1073/pnas.1906589116.

5. James McClintock, "Tips for a Connected Conversation on Climate," Nature Conservancy, December 12, 2018, https://www.nature.org/en-us/what-we-do/our -priorities/tackle-climate-change/climate-change-stories/how-to-have-a-connected -conversation-about-climate-change/.

6. J. Poore and T. Nemecek, "Reducing Food's Environmental Impacts through Producers and Consumers," *Science* 360, no. 6392 (June 1, 2018):991, https://doi.org /10.1126/science.aaq0216.

7. "Do You Consider Yourself to Be a Vegetarian or Vegan?," Statista, released August 2018, https://www.statista.com/statistics/237462/share-of-vegetarians-in-the-united-states/.

8. "Share of People Who Are Vegetarian in Leading Countries in Europe in 2016," Statista, released May 2017, https://www.statista.com/statistics/722215/distribution-of-people-following-a-vegetarian-diet-europe/.

9. Walter Willett et al., "Food in the Anthropocene: The EAT–Lancet Commission on Healthy Diets from Sustainable Food Systems," *Lancet* 393, no. 10170 (February 2, 2019):447–492, https://doi.org/10.1016/S0140-6736(18)31788-4.

10. *Food Wastage Footprint: Impacts on Natural Resources: Summary Report* (Rome: Food and Agriculture Organization of the United Nations, 2013), 6, http://www.fao.org/3/i3347e/i3347e.pdf.

11. Ana Maria Loboguerrero et al., "Food and Earth Systems: Priorities for Climate Change Adaptation and Mitigation for Agriculture and Food Systems," *Sustainability* 11, no. 5 (January 2019):11, https://doi.org/10.3390/su11051372.

12. "Reducing Wasted Food at Home," US Environmental Protection Agency (EPA), accessed March 27, 2020, https://www.epa.gov/recycle/reducing-wasted-food-home.

13. "Food: Too Good to Waste; Implementation Guide and Toolkit," EPA, February 4, 2016, https://www.epa.gov/sustainable-management-food/food-too-good-waste-implementation-guide-and-toolkit.

14. Marco Springmann et al., "Options for Keeping the Food System within Environmental Limits," *Nature* 562, no. 7728 (October 2018):519–25, https://doi.org/10.1038/s41586-018-0594-0; Tristram Stuart, *Waste: Uncovering the Global Food Scandal* (New York: Norton, 2009).

15. Rebecca Smithers, "Raise a Toast! New Beers Made from Leftover Bread Help to Cut Food Waste," *Observer*, April 28, 2018, https://www.theguardian.com/lifeandstyle/2018/apr/28/new-beers-made-from-leftover-bread-marks-and-spencer-adnams.

16. "Inventory of U.S. Greenhouse Gas Emissions and Sinks," EPA, accessed March 27, 2020, https://www.epa.gov/ghgemissions/inventory-us-greenhouse-gas-emissions-and-sinks.

17. "Footprint Calculator," Global Footprint Network, accessed July 11, 2019, https://www.footprintnetwork.org/resources/footprint-calculator/.

18. Project Drawdown, accessed May 15, 2019, https://www.drawdown.org/.

19. Dwight D. Eisenhower, "Address at Bradley University," September 25, 1956, in "Agriculture: A List of Holdings," compiled by Herbert L. Pankratz, February 1991, https://www.eisenhowerlibrary.gov/sites/default/files/research/subject-guides/pdf/agriculture.pdf.

20. Jonathon P. Schuldt, Danielle L. Eiseman, and Michael P. Hoffmann, "Public Concern about Climate Change Impacts on Food Choices: The Interplay of Knowledge and Politics," *Agriculture and Human Values* (January 29, 2020), https://doi.org/10.1007/s10460-020-10019-7.

21. C. Mbow et al., "Food Security," in *Climate Change and Land: An IPCC Special Report on Climate Change, Desertification, Land Degradation, Sustainable Land Management, Food Security, and Greenhouse Gas Fluxes in Terrestrial Ecosystems,* ed. P. R. Shukla et al. (Geneva: Intergovernmental Panel on Climate Change, 2019), 518, https://www.ipcc.ch/site/assets/uploads/sites/4/2020/02/SRCCL-Chapter-5.pdf.
22. Ilona M. Otto et al., "Social Tipping Dynamics for Stabilizing Earth's Climate by 2050," *Proceedings of the National Academy of Sciences* 117, no. 5 (February 4, 2020):2358, https://doi.org/10.1073/pnas.1900577117.

INDEX

US Environmental Protection Agency, 185
US Global Change Research Program, 178

vanilla, 5, 47, 68, 142, 152–54, 162;
adaptive solutions for, 154; effects of
climate change on, 153–54
vegetables, 113, 123–30; effects of
climate change on, 128. *See also*
potatoes
vegetarianism, 184. *See also* plant-based
diet
veggie burgers, 96
Vibrio infections, 105
Vietnam, 19, 101, 116, 156
vineyards, 56–63
vitamins, 31
vodka, 66, 119, 124
volcanoes, 22

Waldorf salad, 70–72
Wales, 125. *See also* United Kingdom
Walmart, 12, 172
walnut trees, 36
Washington State, 52–53, 124–25
waste management, 98
water: avocados and, 74–78; beer
production, 50, 52, 54; contaminated,
110; desalinated, 76–77; effects of
climate change on, 25–30; plant
growth and, 3, 9, 25–30, 45, 115,
118–19; recycled, 76–77; vegetables
and, 128; wheat and, 121. *See also*
floods; glacial meltwaters; irrigation;
oceans

water quality, 170
water shortages, 20, 53, 93; conservation
practices, 54, 126; dairy cows and,
144–45
weather: difference between climate
change and, 18; erratic, 83; in
troposphere, 14. *See also* extreme
weather; precipitation; rainfall
amounts
website for *Our Changing Menu*, 6
weeds, 21, 31, 41, 175, 180–81
Weinstein, Colton, 68
West Virginia, 64
wheat, 113, 119–23, 175; adaptive
solutions for, 122–23; carbon dioxide
and, 31; effects of climate change on,
121–22; temperatures and, 34
whiskey, 64–65, 68, 119
whiteflies, 42–43
wild fisheries, 107, 109
wild rice, 115
wine, 4, 49, 56–63; adaptive solutions
for, 58–63; consumption of, 57; effects
of climate change on, 58–60; history
and types of, 56–57
winter temperatures, 32–34, 53, 83
winter wheat, 121–22
Wisconsin, 142, 144
Worcestershire sauce, 66–67
World Bank, 169
World Cocoa Foundation, 140–41
Wyoming, 92–93

yeast, 49, 51, 53, 57, 181

ABOUT THE AUTHORS

DANIELLE L. EISEMAN is a visiting lecturer in the Department of Communication at Cornell University, where she teaches risk communication, science communication and writing, and environmental communication. Danielle's research examines climate change communication strategies with a particular focus on encouraging pro-environmental behavior change. She also contributes to interdisciplinary research on policy, planning, and resilience at the community level. Her doctorate is in marketing from Heriot Watt University in Edinburgh, Scotland. Danielle also holds master's degrees in carbon management from the University of Edinburgh and in interdisciplinary studies (marketing and economics) from DePaul University, a bachelor's degree in chemistry from Miami University, and a degree in culinary arts from the Scottsdale Culinary Institute.

MICHAEL P. HOFFMANN dedicates all of his time to the grand challenge of climate change and helps people understand and appreciate what is happening through food. Melting glaciers are bad enough, but the loss of coffee is downright terrifying, and this is what keeps him going. He tells the climate change story to a wide range of audiences and has published climate change articles in the popular press: The Hill, *Fortune*, *USA Today*, and Medium. His TEDx Talk—"Climate Change: It's Time to Raise Our Voices"—has been well received. Previous positions he has held include executive director of the Cornell Institute for Climate Change Solutions, director of the Cornell University Agricultural Experiment Station, associate dean of Cornell's College of Agriculture and Life Sciences, associate director of Cornell Cooperative Extension, and director of the New York State Integrated Pest Management Program. He is a professor emeritus in the Cornell Department of Entomology. He received his bachelor's degree from the University of Wisconsin, a master's degree from the University of Arizona, and a doctorate from the University of California, Davis.

CARRIE KOPLINKA-LOEHR learned to haul hay bales and milk cows as a teenager, then spent three decades communicating about agricultural issues and how they intersect with the environment. At Cornell University she directed the Northeastern Integrated Pest Management Center, which continues to fund multistate research and extension projects that reduce environmental risks. She also developed and led the communications team for the New York State Integrated Pest Management Program in Geneva, New York. As a freelance writer, Carrie has published in *YES!* magazine, *Sierra* magazine, and other venues. She lives in a student-designed solar home and is an avid gardener and land conservationist. With a master's degree in science education from Cornell University and a bachelor's degree in English from Colgate University, Carrie is a member of the Society of Environmental Journalists and the National Association of Science Writers.